How to Take Pictures of an Eclipse

of an Eclipse

An astrophotography beginner's guide to capturing solar and lunar eclipses

Allan Hall

Acknowledgements:

As always I have to thank my wife, Sue Ann, because without her prodding and help, my books would either never get done or would be far less than they are.

I would also like to thank my readers. Their emails, letters and reviews let me know that my work is appreciated which drives my desire to create more and better books.

Table of Contents

1: Introduction

I have been fortunate in my life to observe several lunar eclipses and one annular solar eclipse. By observe I mean that I watched the entire event in awe. I have also seen one total solar eclipse however I was so young I do not remember anything about it other than it let me get out of class for half a period. I did not appreciate it and I deeply regret that.

I am writing this book as I prepare for the total solar eclipse in 2017 which will be over forty years from the last total solar eclipse I was around for. They don't come often and I will be prepared for this one.

Typically I am just like any other astrophotographer, taking images of objects that move very little. Sure they move across the sky, but my average exposure time on a single object is usually somewhere between five and thirty minutes per image. Then I take between ten and fifty of those images and combine them into a single picture. Eclipses are wildly different although I can use the same equipment.

The eclipse and transit (when something apparently much smaller like the planet Venus moves in front of something that appears much larger like the sun) are the only times we astrophotographers get to photograph objects really moving and changing in minutes, not years or even millions of years.

Taking pictures of a solar or lunar eclipse can be a lot of fun and easy to do. There are many considerations however and since you might be taking pictures of something that you may not get the opportunity to do again in your lifetime, it will help greatly to do a little research before the event.

While a lot of equipment can be used for this task, you certainly do not have to have much just to take some fun pictures. In fact, some amazing images can be taken with what you most likely already have right now.

Of course there are things you can acquire to give you more options should you choose to go that route. We will look at some of those too.

This book aims to show you what you can do with what you have, as well as give you ideas on what you could do with more or different equipment. It is aimed more towards the newcomer to this type of thing and not the seasoned astrophotographer, although both will surely gain something.

Those of you who have already read my other books such as *Getting Started: Long Exposure Astrophotography*, *Getting Started: Budget Astrophotography* or *Getting Started: Visual Astronomy* will notice that some of the information in those books is repeated here. This only makes sense as taking a picture of an eclipse is both astrophotography and astronomy.

What this book aims to do differently is to expand on the ways in which you can image an eclipse, and provide more information on the equipment and techniques specific to eclipses. You will not find information on imaging deep space objects here, and you will not find information on changing exposure settings as an eclipse progresses in those other books.

My hope is that whether you snap a quick image with your phone through a piece of solar film or buy a solar telescope complete with computerized mount and high frame rate video camera, you have fun. If this book helps you even a little in that endeavor then I am a happy person.

2: Warnings

It is difficult for me to overemphasize the danger in looking directly at the sun, or for that matter, pointing any optical device at the sun without the proper filters.

Taking a quick glance at the sun might just render you temporarily blind or make you see spots, or it could also do permanent damage. Either way, it is very bad.

If you think that is bad, I am sure you have heard of the little kid using a magnifying glass to burn things with only sunlight. A camera lens, telescope, or pair of binoculars could potentially amplify the sunlight far more than a simple magnifying glass and therefore will possibly have no problem setting things on fire, including your eye.

Let's be clear, never under any circumstances look at the sun without a real solar filter. Some people will tell you that welding glass will work too, maybe it will, but I am not willing to take that chance and urge you not to either.

I have also heard people talk about using spray tint, multiple layers of window tint, and "really dark sunglasses". No, no, and oh heck no.

Never point a camera at the sun without a proper solar filter either. Even if you are not looking through a viewfinder the amplified sunlight can burn the sensor and possibly even set it on fire.

Just don't!

3: The basics

Before we get into much about equipment we need to cover some of the basics so we understand not only what we want, but how to achieve it.

The following sections start with defining an eclipse (at least for our purposes) and attempt to get the answers we need to move forward.

3.1: What is an eclipse?

An eclipse is simply where one celestial body (for example the moon) passes in front of and obscures a portion (or all) of another celestial object (such as the sun).

If the moon passes between you and the sun, that is a solar eclipse (as the sun is being eclipsed by the moon). If the Earth passes between the sun and the moon while you are viewing the moon, that is a lunar eclipse.

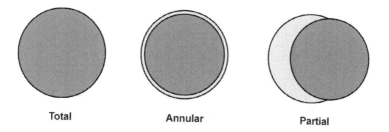

Total   Annular   Partial

There are several different types of eclipses including partial, annular and total.

Partial eclipse means that only a part of the object is covered. For example, a partial solar eclipse means that only a portion of the sun is covered by the moon.

An annular eclipse is really a partial eclipse where the object in front does not completely cover the object in back however it does travel directly through the center. The point at which the moon is perfectly centered in the middle of the sun is called annularity.

A total eclipse means that the object in front completely covers the surface of the object in the rear. A total solar eclipse for example is where the entire surface of the sun is visually covered by the moon. The point at which the sun is completely covered by the moon is called totality.

In a total eclipse we talk about the surface being covered because the corona (whisks of solar radiance emanating from the surface outward into space) can be seen apparently radiating out from the surface of the moon.

Eclipses are defined as partial, annular, or total and independent from a particular viewpoint. This means that if an eclipse is a total eclipse from anywhere on the surface of the Earth, it is considered a total eclipse. Even if there is no one at that point on the planet (such as the North or South Poles, in the middle of an ocean, or in an uninhabited desert, it is still a total eclipse.

If you view a total eclipse from an area where the eclipse appears as a partial eclipse that does not change the designation of the eclipse type from a total eclipse.

If a much smaller object (visually from Earth) passes between the Earth and another object such as the sun, that is called a transit and not an eclipse.

2012 was an interesting year as it included both a transit of Venus (Venus appearing to cross the surface of the sun) and an annular solar eclipse in the United States.

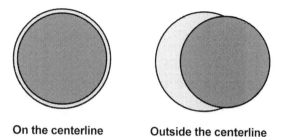

On the centerline   **Outside the centerline**

The best place to view an eclipse is typically on a specific path called the centerline. This is the path on the Earth directly below the event. This path is where you will see the moon totally centered in the middle of the sun during an annular eclipse or the moon completely covering the sun during a total eclipse.

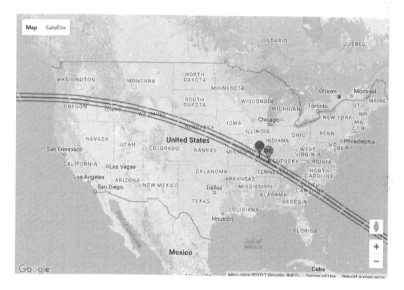

The map above shows the centerline of the 2017 total solar eclipse. Look closely and you will see three lines, two outside and one in the center. The line in the middle is the exact centerline and where you want to be if possible.

7

Anywhere between the two outside lines will still give you a complete total eclipse but it may not be as exact as it could be if you are on the outside areas.

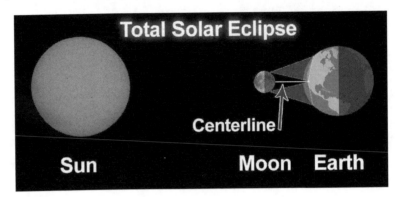

The image above shows how a solar eclipse looks from space including where the centerline is. Note that the illustration is not to scale.

If you are off the centerline a substantial amount (way outside the shown area) you may not see the eclipse happen at all!

The two markers in the map above are GD standing for Greatest Duration of the eclipse and GE which stands for Greatest Eclipse. If you want to get the absolute best possible view of the eclipse, you need to be on the centerline at those markers.

There are five stages to a total or annular solar eclipse: first contact, second contact, totality (or annularity), third contact and fourth contact.

First contact is when the moon just touches the outside edge of the sun.

Second contact is the moment the moon completely covers the sun.

Totality or annularity is everything starting with second contact and ending with third contact.

Third contact is where the sun just starts to peek around the opposite side of the moon.

Fourth contact is when the sun is no longer covered at all by the moon.

Partial solar eclipses have three stages: eclipse begins, maximum eclipse and eclipse ends.

More information including interactive maps for solar & lunar eclipses as well as transits can be found at NASA's website here:

https://eclipse.gsfc.nasa.gov/

3.2: Finding a place to take your pictures

Depending on the rarity of the eclipse you are looking to image, finding a spot to set up can be a difficult thing. Lunar eclipses are generally far easier as they happen more often and have fewer people looking to view it. Full solar eclipses can be extremely hard to find a good place that isn't packed with people.

During the 2012 annular eclipse my wife and I traveled to Albuquerque, New Mexico from Huntsville, Texas as it was on the centerline. It was also closer to the west coast which meant that the eclipse would occur higher in the sky.

Since the objects in the sky (sun, moon, etc.) rise in the east and set in the west you need to take this into consideration in your plans. If the sun sets at 7pm on the east coast and the eclipse reaches totality at 7:15pm you will miss it. Drive an hour west and you would be able to see totality.

Now that we know that moving either east or west will affect the altitude at which the eclipse happens we need to decide what altitude we want to image the eclipse. You want the eclipse to occur above the horizon but below the zenith (the point directly overhead).

The reasons for this is that too close to the horizon means there is substantially more atmosphere you have to see through and the pollution and water vapor in the atmosphere distort the view, making everything less sharp.

Shooting directly overhead is good, except it is the most difficult to do. Equatorial mounts tend to have problems shooting directly overhead requiring some yoga moves called a meridian flip. Other telescope mounts have difficulty with camera attachments at zenith. Not to mention the crick in your neck you get from prolonged looking straight up.

The best of both worlds would be for the event to happen somewhere around 45-60 degrees above the horizon. Try to pick your location to get you in this range.

Albuquerque was a full day's drive in a car packed with astronomy gear. We stayed at a hotel in town which we reserved several months in advance.

Since this was "just" an annular eclipse there was a smaller number of people traveling to the centerline than there would have been had it been a full eclipse. Even so, most of the hotel rooms in the Albuquerque area were booked solid.

On the day of the eclipse, there were no less than three large organized viewing locations in the area. We picked the one nearest the airport on the south side of town as it seemed to be the best organized and was in a massive parking lot which would make setup easier since it was all paved.

In the above image you can see one of the groups of people at the 2012 annular solar eclipse in Albuquerque who wanted to set up on the grass instead of on the asphalt like all the astrophotographers, including myself)

My best guess was that there may have been a couple thousand people at our location alone. Add a few thousand at the other locations in the area and there may have been somewhere between five to ten thousand people viewing the eclipse in organized areas around that one town.

That does not include all the people viewing from other places such as their homes, schools, or just stepping out of their offices for a few minutes.

Remember, that was "just" an annular eclipse. At a full solar eclipse you can expect far more.

Assuming you do not want to be at an organized group location, and assuming you have looked up where the centerline of the eclipse will be and decided the altitude you want the eclipse to happen at, the next step is looking for unique places.

Some great places can be rest stops, scenic overlooks, large parking lots for malls, the rooftop level of a multilevel parking lot, or the rooftop of any tall building.

I shot the transit of Venus in 2012 at a rest stop with a couple of other people who found the same rest stop I did.

When looking for something like this you first need to find the centerline and get a rough idea of where you want to be. Open Google Maps and turn on satellite view. Now go to the area you want to be in and start looking around.

Start with a wide view looking for large parking lots, cleared dirt areas, rest areas, etc. Note in the previous image I found what looked like a cleared area just off the freeway's service road. It was in an excellent spot as there are no gas stations or other businesses in the area, cutting down on traffic. I could also look out over a pasture with no apparent trees to the west (which is where my target would be).

An ideal area would be near a cliff or on the edge of a lake. Of course make sure your back is not to the cliff or lake as that would ruin the advantage.

Zoom in a little to make sure it is what you think it is, and that there is access directly from the road without having to drive across a pasture (ranchers hate that for some reason).

This spot seems accessible from two different roads and is fairly large.

Now zoom in to street view and make sure the area is indeed flat and in good shape, that there is no fence or other obstruction you have to deal with and most importantly, that there is a good, clear, unobstructed view in the direction the eclipse will happen.

As long as the property you are planning on viewing from is not private property and you are doing a solar eclipse in the daylight, you are unlikely to be bothered.

You will be amazed at all the places you find using this method. I would advise that you find three or four places in close proximity in case the satellite images are old and do not represent the current state of the place. The cleared place may be filled with road construction equipment, flooded from rain, overgrown, or fenced off. Having several backups makes sure you are covered no matter what happens.

If at all possible, visit one or more of your locations beforehand and know what to expect.

For lunar eclipses you need to be away from headlights and bright lights. While imaging a lunar eclipse does not need darkness like long exposure astrophotography does, it still will benefit from not being in the middle of town.

16

Contact your local high school or university and see if they are planning something for the event. Some universities may even have an observatory which allows the public at events like a lunar eclipse.

If not, try a parking lot for a mall or large department store on the outskirts of town. Remember this is an event that takes place at night so make sure you are somewhere safe, preferably in a group.

My last tip in this section is to leave early. It would really stink if you got stuck behind a wreck, a funeral procession, or in road construction and missed the eclipse. It is much easier to sit at your spot and read a book for half an hour waiting for it than in traffic screaming at the situation.

3.3: The difference in photographing solar and lunar eclipses

There are surprisingly a lot of similarities between the two. The main differences are that you have a much harder time getting the sun in your camera frame than you might think since you cannot really look at the sun. The moon on the other hand is simple since you can look right at it.

The sun requires filters, the moon does not. You may however want to use lunar filters when viewing the moon, or even when imaging it. Remember that the moon is just a light colored rock in direct sunlight for at least a portion of the time you are watching it. This means that your exposures would be similar to if you were photographing a rock on the ground during the day.

Lunar eclipses require constant fiddling with the exposure as the eclipse progresses, solar eclipses do not. Total solar eclipses however can let you take off the solar filter at the moment of totality to take a few images before reinstalling the filter.

3.4: Filters for use with solar eclipses

There are two basic types of filters you need to consider if you are not going to use a dedicated solar telescope (which very few of us own or can reasonably afford): factory made and homemade.

Factory made comes in two basic varieties: glass and film.

I prefer glass filters as they in my opinion seem to produce sharper images, are more rugged (don't crease or crinkle if you touch them wrong) and are not susceptible to pinholes.

A pinhole in a filter can allow a bright point of light in that at best can ruin your pictures and at worst, blind you.

Film filters are less expensive and tend to be available for a wider array of telescopes, cameras, and other devices. In all fairness I have seen no real evidence of a pinhole problem although I have heard it can happen.

Homemade filters are almost always made from film. You purchase a sheet of film such as Baader solar film and use that to cover whatever you plan on pointing at the sun. You can make glasses out of it, use it to create a screw-on camera filter (as shown later in this book), and anything else you can think of.

I use it to create filters for binoculars, solar viewers (a piece of cardboard with a solar film window in it as shown in the previous image), and even to cover video camera and smartphone lenses.

The best advice I can give you is to always have spare solar film on hand in case you lose or break your glass filter, and also to cover devices you didn't think to get filter for (or could not find one premade).

I also urge you to order your filter(s) as soon as possible. If you wait until the month before the eclipse, no one will have any in stock.

I ordered my primary telescope's glass filter six months before the eclipse and almost didn't get the one I wanted as supplies were already starting to dry up.

I ordered the last batch of solar film about two months before the eclipse and had a devil of a time finding anyone with any in stock. This time I have all I need and will be ordering some spare about six months before the big 2017 eclipse just in case.

For glass filters I prefer Thousand Oaks Optical at www.thousandoaksoptical.com. I have had great luck out of their filters and will not hesitate to purchase new ones there if I need one for my primary telescope.

Although Thousand Oaks Optical sells highly rated film as well, Baader solar film is the film most people turn to. It is cheaper but less robust. In most cases people purchase the film for one time use which makes Baader a better choice.

Some cameras can be difficult to put a solar filter on because the lens retracts into the body of the camera and may even be covered by a little door when not in use. This prohibits the use of a screw in filter, or even taping a film filter over the lens. The solution for this is to make a film card and hold that in front of the camera while using it.

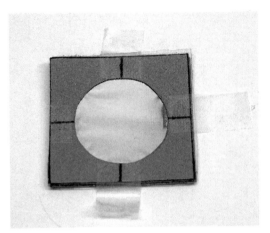

These cards are easy to make with just a couple pieces of cardboard, some tape, and the solar film. Just cut a hole in two equal sized pieces of cardboard, cut a piece of film larger than the hole in the cardboard, tape the film to one piece of cardboard and then tape the other piece of cardboard over the film and first piece of cardboard to "seal in" the film.

Don't worry about imperfections in the film such as a crinkle or crease as long as the imperfection does not allow any light to pass through without being appropriately filtered.

This method prevents the film from snagging on something, or being blown loose by wind, causing all kinds of really bad things to happen.

If however you want film for something more permanent, or a better quality optical film, Thousand Oaks Optical is the way to go.

When photographing a total solar eclipse, when the eclipse reaches totality, you can remove the solar filter and image the eclipse directly.

The above image is an example of what you can capture at the moment of totality during a total solar eclipse with the solar filter removed.

Be sure you are careful and reinstall the filter immediately when the sun starts peeking around the edge of the moon again.

You may run into telescopes which use filters over eyepieces, or special eyepieces for solar viewing. I highly suggest you throw those away and use a filter over the front of the telescope instead. The reasoning is purely safety related.

Sunlight going through a telescope at full strength tends to heat up the telescope which can cause things to fail. In addition, when the light passes through the front objective of the telescope (or through the mirror assemblies) it is magnified and concentrated meaning any failure of a solar filter on the rear of the telescope would subject your eye or camera to super concentrated sunlight, destroying your eye almost instantly and damaging your camera as well.

3.5: Exposure settings

The ideal solution to taking pictures of an eclipse is to use a camera where you can vary the exposure. With a DSLR or high end point & shoot, that can be accomplished by changing the shutter speed, aperture of the lens, or the ISO (sensitivity).

The reason this is so important is that as you go from a fully exposed moon to it being even partially eclipsed, it gets darker. A lot darker. The reason for this is the light that is illuminating the surface of the moon goes from traveling directly from the sun to the surface of the moon, to passing through the atmosphere of the Earth. This substantially reduces the brightness of the light making the surface dimmer.

This light will also slowly turn from white to a red/orange color as the eclipse approaches totality (where the moon is completely behind the Earth).

The previous image of the moon was at the very start of a full Lunar eclipse and was shot with a Nikon D7000 on a refractor telescope with an exposure of 1/1000 sec at ISO 200.

Half way through the eclipse the exposure dropped to 1/60 sec at ISO 800.

By the end of the eclipse I was shooting 1 sec at ISO 800 to get the image above.

In the middle of a total solar eclipse when the sun is completely blocked by the moon you may also want to increase your exposure to capture the prominences protruding from behind the moon.

A partial or annular solar eclipse really does not suffer from this issue as at least part of the sun is always directly in the view of the camera and is always the same brightness (the moon has no atmosphere to distort the light passing near it).

Auto exposure settings will not adequately account for this since there will be extremely bright spots in the image along with comparatively very dark spots.

You need to be prepared to make these changes extremely quickly. The eclipse will not wait for you to get it right. When in doubt take more pictures with a range of exposure values because odds are, you are not going to get to reshoot the eclipse any time soon.

Exposure settings vary greatly depending on the filter you are using, the amount of magnification you are using (focal length of the camera lens or telescope), etc. I can however give you some guidance by telling you what my exposures were in shooting a couple of eclipses. You can use this as a starting point and vary it from there to match whatever equipment you are using.

With my 110mm f7 770mm focal length refractor telescope and a Thousand Oaks Optical glass filter I was shooting 1/800sec at ISO 400 using a Nikon D7000 DSLR.

A second camera, a Nikon D90, was using a 300mm lens shooting through a Baader film solar filter at 1/1000 sec, ISO 400 using f16.

Shooting during totality of a total solar eclipse can have a wild variance of exposure settings depending on what you want to capture (how much of the prominences, the amount of detail you want to get out of them, etc.), and the relative size of the moon at the time (the moon is sometimes closer to the Earth than others).

Your best bet when shooting during totality is to shoot a ton of images in a wide range of exposures and hope for the best. Since digital images can be manipulated in software after the fact, be sure to shoot in RAW mode and vary your exposures by at least two stops if not four. This means if you shoot your first exposure at 1/500 sec, shoot the next two at 1/125 sec and 1/2000 sec.

3.6: Focal lengths / magnification

A common question is, "what will the sun/moon look like in my camera?" Or "what lens do I need to buy to get a decent image of the sun/moon?"

There are two things to take into consideration when answering this question: focal length of the lens/telescope, and the size of the camera sensor. Let's look at how this works.

Lenses and telescopes do not actually magnify an image, they present a smaller amount of an area to the camera sensor, or your eye. This is called increasing or decreasing the field of view. A wider field of view contains more of the scene (lower apparent magnification) whereas a narrower field of view contains less of it (higher apparent magnification).

To give yourself an example of this cut a one inch square in a piece of paper and hold it about a foot in front of your face and look at a distant object. Now hold that same piece of paper at arm's length and look through it again.

You should see the exact same scene only a smaller portion of it. That is the basic idea of how a lens or telescope works.

The projected image from the lens falls onto the sensor of your camera. The larger the camera sensor, the larger the field of view. The smaller the sensor, the smaller the field of view.

Take a DSLR camera for example which typically comes with one of two sensor sizes: full frame and crop sensor (or APS-C). A full frame usually provides higher quality images, particularly in low light. A crop sensor provides a smaller field of view and a smaller, lighter package.

Crop sensor cameras such as those from Nikon typically provide a 50% increase in apparent magnification. This means that if you are shooting with a 100mm lens on a crop it will provide the same picture as shooting with a 150mm lens on a full frame camera (quality and low light capability aside).

Let's take a look at what kind of images some of my equipment can deliver so you have an idea what to expect. Each of the following images have not been cropped so you can see how large the target is in the final image.

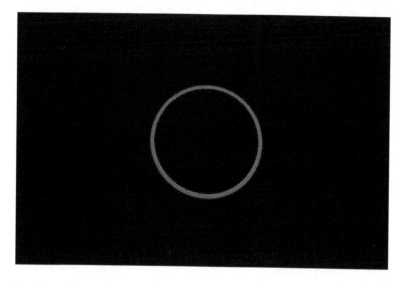

The above image was shot with a Nikon D7000 APS-C crop sensor camera without a lens, directly connected to a 700mm refractor telescope. The image is uncropped.

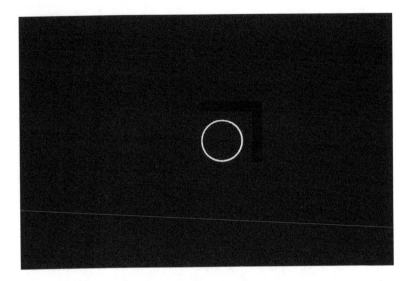

The previous image is of the same eclipse, at the same time, with a Nikon D90 camera using a 300mm lens.

Above is an image of the moon shot with a Nikon D7000 on a 700mm refractor telescope.

The image above was shot with a Nikon D7000 and a 200mm lens and is of a super moon (when the moon appears the largest in the sky).

Now don't get discouraged with the smallness of the previous image. All the images so far have been to show you how big the sun or moon was in your viewfinder but remember you can crop the image once you have the photos taken.

Above is the same image as the previous one except this one is cropped. The higher resolution camera you have, the more you can crop the image.

Most DSLRs either come with a lens that zooms to 200mm or you have access to one from a friend so this does not require any specialized equipment.

Approximate size of an eclipse at a given focal length for a full frame camera

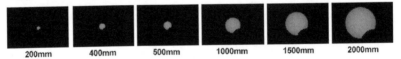

| 200mm | 400mm | 500mm | 1000mm | 1500mm | 2000mm |

The previous chart will give you an idea of what your camera and lens combination can do. Keep in mind that a crop sensor / APS-C camera with a 200mm lens is the equivalent of a 300mm full frame.

Also remember that you can crop the image after shooting which works particularly well with high megapixel cameras.

Another thing to keep in mind is that the sun and moon are roughly the same size when imaging.

Now let's assume that you have a DSLR and lens, or a DSLR that you are connecting to a telescope and you want a little more magnification without breaking the bank, what can you do?

If you are using a DSLR with a lens, you can use a teleconverter. This device is a small lens that mounts between your camera and lens which can effectively magnify the image 1.5x, 2x or even 3x.

Teleconverters can be had starting at around $100 for a 1.4x or 1.5x off brand model (Vivitar and Kenko are popular choices) up to $400 or $500 for the Nikon or Canon models.

The name brand ones are usually substantially better quality optics and allow for auto focus, whereas the less expensive off brand ones are usually inferior in quality and features.

The lower magnification, the less degradation to the image as well so your best bet is a 1.4x-1.5x model from Nikon or Canon.

Realistically however, the 1.4x or 1.5x off brand models provide more than adequate image quality at a very attractive price.

A barlow is the same idea as a teleconverter, but for a telescope. It goes between your eyepiece or camera and the telescope and comes in some of the same magnifications such as 2x and 3x as a general rule.

The same advice applies to barlows as to teleconverters however barlows are generally much cheaper ranging from $50 to the ultimate Teleview Power Mate models commanding almost $400.

Again, stay at the lower magnifications around 2x and stay away from the bottom (the cheapest you can find). Midrange models of just under $100 for a 2x model provide excellent results.

3.7: Methods of firing the shutter

Although we discuss the intervalometer and software to fire the shutter a little later in the section on cameras, we need to delve a little more into that now, primarily to understand why it is so important.

When we take astrophotos, especially ones of thirty seconds or less, it is important to understand that every time the shutter fires, the camera vibrates. While this is not good, and there are things we can do to lessen its effects such as making sure the mount is as balanced and stable as possible, it pales in comparison to the vibration caused by you touching the camera to fire the shutter.

No matter how carefully or lightly you do it, the image will suffer. I know what you are thinking, not you, you can be really careful and not touch anything and you have really light fingers. Sure you do....

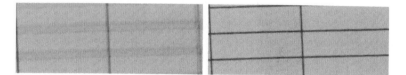

The previous image shows an enlarged crop of the same target (a grid), taken by the same camera, with the same settings, seconds apart. The difference is the one on the left was gently pressed by hand while the one on the right was fired with a remote control so nothing was touching the camera. Big difference!

For this test the camera was mounted onto a professional Manfrotto carbon fiber tripod with a Manfrotto magnesium pan/tilt head, about $450 worth of professional level support. Can you imagine what would have happened had I put it on a cheap $50 tripod instead and then pressed the shutter release button?

Some astrophotographers will tell you that remotes don't matter, and that shutter vibration doesn't matter, and they may be right. That is if they are talking long exposure astrophotography instead of the short exposures we are typically doing while shooting eclipses. The reason is simple, it may take up to five seconds for the vibrations to stop, and if we are taking a five second exposure that is 100% of the time it is vibrating, that's a big problem. With long exposures however, that five seconds of vibration may be far less than 1% of the total exposure time, and therefore completely insignificant.

So now we have to decide how to minimize these vibrations. One trick if you are using a DSLR is to see if your camera has Mirror Lock Up. This feature takes the mirror that reflects the image from the lens up through the viewfinder and locks it in the up position the first time you release the shutter, then the second time you fire the shutter release it actually releases the shutter. Since the mirror slapping up is a good part of the vibration on a DSLR this can really help reduce vibrations.

The next thing is you can use the built in timer in most cameras, which on some DSLRs will also lock up the mirror when you start the timer. This allows you to be away from the camera with any vibrations you may have caused long settled before the shutter is opened.

Of course you can use the intervalometer we discuss later which has the added benefit of allowing you to set up a lot of exposures all in one run without you having to babysit and continuously fire off the shutter. The intervalometer runs off two AA or AAA batteries so you don't need an outlet or large battery pack, and of course the whole thing fits into your pocket. I personally believe this is the best bang-for-the-buck solution for shutter release in budget astrophotography.

Lastly, you can use a computer to release the shutter using software such as DSLR Shutter from Stark Labs, available from:

http://www.stark-labs.com/page26/DSLR_Shutter.html

The advantage to software such as DSLR shutter is if you already have a computer and cable for the camera, this solution is free and very easy. The down side is having to lug a computer into the field all the time.

3.8: Camera stabilization

Very little is more important than a stable camera. Without the stability, how the camera attaches, whether it tracks or not, how much magnification you have, what resolution your camera is, all means nothing. Unless of course you want to take pictures of blurry little objects; do you? I thought not.

Stabilizing your camera can substantially increase the sharpness and definition of your picture. A lower resolution picture that was taken on a tripod almost always looks better than a high resolution image that was shot hand-held.

Now stability comes in many shapes and sizes and does not necessarily require expenditures of vast quantities of money. Most of the time making something stable requires more thought and common sense than anything else.

I have a pretty substantial mount for my main astrophotography rig, and it was all balanced and running smoothly one night. The entire thing was sitting on a large concrete slab a couple feet thick, very stable. Then this 150 pound kid started jumping up and down on the concrete a couple feet away.

Was my mount stable? Yep. Was the ground stable? Yep. Did this kid cause vibrations that my scope could see? Yep. And if I could have figured out a place to bury.....oops, forget I said that, but this is Texas you know ☺

I politely asked the kid to stop and explained to him why he should not do that ever again.

The most basic way of stabilization is setting your camera on a table, bench, top of a car, etc. This can work surprisingly well, especially when shooting eclipses. If you can, use something to support the camera like a bean bag, wadded up shirt or a jacket. Anything to support the camera is better than trying to hold it in your hand.

If you must hand-hold your camera or phone, brace yourself or your camera against something, anything, a tree, the side of a building, telephone pole, signpost, whatever is handy. Control your breathing and click the shutter right in the middle of a slow exhale.

I have seen people use fence posts or other poles to place their camera on to take images, and while this will work, be sure the camera is secured so that it cannot fall to the ground. No image is worth a broken camera.

The next best thing of course is a tripod. Here, like in most things, you get what you pay for to a point. Fortunately, you don't need a lot to hold just a camera but you don't want the cheapest tripod you can get your hands on either. Splurge just a little and get one slightly beefier than you think you will need. Remember that when you click the shutter, that causes vibrations and the heavier duty the tripod, the less the vibrations will matter.

You might be able to even use a tabletop tripod for a small camera. This would be far more portable, and can be more stable in some cases. You do however have to watch out and keep yourself or anyone else from touching the table, and also note that wind can play a significant role as it can wobble tables as well.

One thing no one thinks about is ground vibrations. If you set up your tripod on someone's wooden deck, and then someone walks across the other side of the deck, those vibrations will travel through the deck and right up the tripod. Keep the tripod on solid ground and away from any vibrations (or children jumping up and down).

One trick when using a tripod is to never extend it to its maximum height as this is when the tripod is its weakest. Try to keep it to 75% or less of its maximum extension. On my primary imaging mount I never extend the tripod at all even though this makes the telescope far too close to the ground to look through, my camera does not have to bend over to see through it ☺

Use buildings or walls as wind breaks to keep the wind from moving your camera. Never touch the tripod or camera while you are taking an image, and always wait a minute or two after touching either before firing off another picture.

You can also generally purchase tripods used as they tend to be pretty resilient. This allows you to buy a nice, heavy tripod for the money of a cheap flimsy one.

3.9: Non-photographic considerations

Photographing a solar eclipse will set you outside in bright sunlight for an extended period of time. Many people don't really think about things such as a hat, sunscreen, and drinking lots of liquids but these can be very important.

Even if you are out in cooler weather you can get some serious sunburn, particularly because you are looking in the direction of the sun which is not normal for most people.

Make sure you use sunscreen with a high SPF rating and apply it liberally and often. I also suggest taking a cooler with some cold drinks if it will be warm outside. A cool drink of water makes the weather much more bearable.

Bring some white towels or rags and put them over your camera equipment to help keep them cool. This is particularly important if you use a black camera or black binoculars. It really hurts when you pick up those black metal binoculars that have been sitting in direct sunlight for half an hour!

For lunar eclipses you may be outside in the middle of the night not moving for quite some time. If the temperature falls below about seventy degrees Fahrenheit then you might start to get pretty chilled. Without much movement over a long period it will start to feel far colder than the thermometer displays. Be sure to take jackets and maybe some blankets.

It is far more pleasant to take a jacket off or leave the blanket in the car than it is to freeze your rear end off waiting for the eclipse. Not to mention you will enjoy the eclipse more if you are comfortable.

4: Determining what kind of images you want

There are many ways to take pictures, or videos, of an eclipse. The last solar eclipse I imaged I used one DSLR camera looking through a telescope which tracked the sun providing a high magnification image, another DSLR attached to outside of the telescope with a 300mm lens and a video camera also attached to the telescope for tracking.

All three of these cameras gave me different images. Since I am not rich enough to travel the world and image solar eclipses I have to do as much as I can when I get the chance. The upcoming total eclipse in 2017 I will probably have at least six cameras going at once time.

Most people however will have much simpler needs. I will start the conversation off with an extreme example costing thousands of dollars and get simpler as we continue. Find what works for you.

4.1: Serious imaging with a dedicated solar telescope

There are many telescopes out there designed specifically for solar astrophotography. Typical models of these are ones that view only a tiny portion of the visible light spectrum called Hydrogen Alpha. These are far superior to just using a filter over a standard telescope and allows them to capture incredible detail. It is the same sort of technology (albeit on a much smaller scale) as what NASA uses to make images such as this one:

While you are not likely to be able to afford telescopes that can generate images like that, you very well could afford a Coronado PST telescope which can create images such as the following:

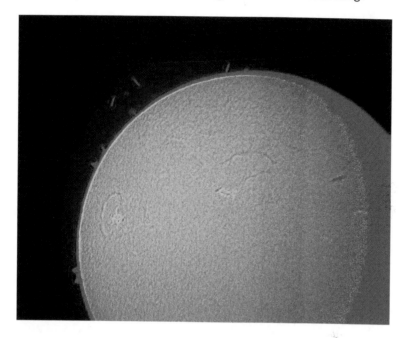

The previous image was taken with a Coronado PST (Personal Solar Telescope) using eyepiece projection into a Nikon DSLR.

These Coronado telescopes start at around $600 and go up from there. Really nice Lundt solar telescopes can start at around $1,000 and easily exceed $4,500. Keep in mind that these prices are only for the telescope tube, no mount, eyepieces, cameras, or adapters.

4.2: High resolution imaging with a serious telescope

You may already be an amateur astronomer and have a nice telescope. With that and an appropriate solar filter you can generate some nice images.

The above image was taken with a standard telescope, Nikon D7000 DSLR camera, and a glass solar filter. What you see in the image is the sun complete with some sunspots, the planet Venus (the round dot on the right center), and an airplane at the top.

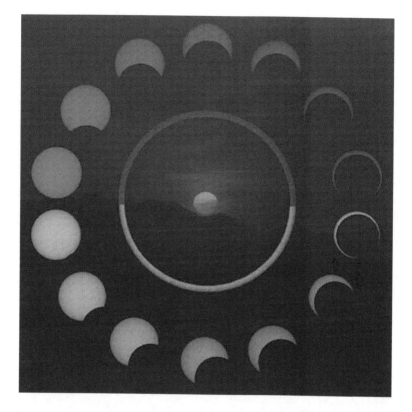

The above image is actually a compilation of images from the 2012 annular solar eclipse. The top (Orange in original photo) images were taken through a glass filter while the lower (gray in the original photo) images were taken through solar film. The images range from just as the eclipse started on the left side to almost a complete eclipse on the right, with the total annular eclipse in the center.

This setup is also very good for lunar eclipses such as this:

There are three huge advantages to this type of setup: stability, tracking, and optical quality.

Here you can see my setup which I used at both the eclipse in 2012 and Venus transit. Note there are two DSLRs: one on the telescope and one on a tripod on the other side. If you look really close you may notice there is a video camera bolted to the side of the telescope as well.

In my setup the primary camera (and video camera) on the telescope are tracking the action so I do not need to do anything except sit back and watch. The telescope is tracking the sun by itself and I am looking at the images as they are taken on the laptop screen under the cardboard hutch on the right.

I actually use this "free time" to manually adjust the second DSLR on the tripod and to use the binoculars to get a firsthand look while the cameras are doing their thing.

If you already have a telescope that you can attach your camera to, or you want to make the absolute most of the experience, then a quality tracking telescope and mount combination is absolutely the way to go.

If not, then you certainly do not have to spend a fortune to get some great images that you will cherish for the rest of your life.

Read on....

4.3: High quality images with only a DSLR

A DSLR can be an excellent way to capture images of an eclipse, either solar or lunar. The biggest issue most people have is that they do not have a "long" enough lens, meaning enough magnification.

With today's high resolution DSLR camera you can make impressive images with a 200mm or 300mm lens and an APS-C or crop sensor size camera. This brings your effective focal length to around 300mm or 450mm respectively which as you will see later is more than sufficient to capture excellent images of an eclipse.

In fact, even when I am using my thousands of dollars' worth of telescopes, mounts, and DSLR camera gear I always have a second DSLR running as well as a backup. This backup camera has produced some excellent images.

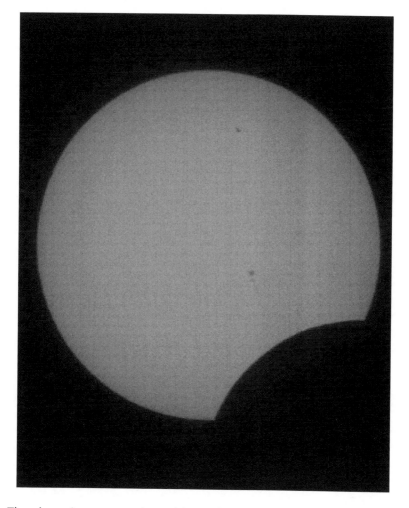

The above image was shot with an older Nikon D90, an inexpensive 70-300mm G series lens and a homemade solar filter using Baader solar film.

This picture is of the 2012 transit of Venus using the same camera, lens, and filter.

The trick is to make the camera as stable as possible, for example using a tripod, and using either the built-in timer or a remote release device to minimize vibrations.

4.4: Good images with less expensive dedicated cameras

Many people still have other types of cameras including point & shoot, bridge, and video cameras. While we cover these in more detail in future sections, they are fine for taking some images as well.

Keep in mind that they will not be of the quality of a DSLR with or without a telescope (the larger sensor size in DSLRs provide superior image quality, not to mention all the other benefits). That may not be what you are worried about.

Most of these types of cameras are more than capable of providing awe inspiring images with a good filter, tripod, and some form of timer or remote release.

Fortunately, virtually all dedicated cameras have a tripod socket on them and some form of timer.

4.5: Quick images with your phone or tablet

I once heard a saying that went something like this: "The best camera in the world is the one you have with you." I have to admit, I have found that true on more occasions than I can count.

If you are like me, your phone and possibly tablet are constant companions. While they will not take exceptional images of an eclipse, they work far better than nothing.

Just like other cameras the best images can be obtained by having an adapter that mounts your device to a tripod. Once the device is stable, using the built in self-timer further reduces vibration and provides a sharper image.

You can tape a piece of solar film directly over the camera to provide protection from the sun's rays (and so you can actually get an image instead of a solid white picture). Be sure that you tape the film securely because if the wind blows it off the lens, you will certainly ruin the image and possibly more.

The above image is solar film taped over the camera of an iPhone.

Now take a piece of cardboard about the size of a piece of paper and put a hole near the center of it.

Hold the cardboard up so the camera shoots through it, this protects the back of the camera and you while pointing towards the sun.

4.6: Fun images with any camera and a pinhole viewer

There is an easy and safe way to not only view, but take interesting pictures of a solar eclipse with very little effort or money.

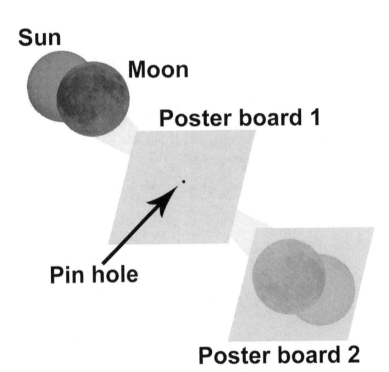

Take two pieces of fairly heavy poster board (or anything similar such as old yard signs, ¼" plywood pieces, etc.) roughly the same size. Place one on the ground at an angle towards the solar eclipse. Make a very small hole (like from a needle or tack) in the other board right in the center and hold it between the eclipse and the second poster board on the ground.

This will project an image of the eclipse onto the poster board on the ground that you can watch and photograph. You can attach the board with the hole in it to a tripod, the side of your house, or anything that will hold it in place for you.

You can spice this up by using something interesting for the board on the ground such as something with a pattern, a different color (although anything other than a very light color will make it hard to view once the eclipse is in totality.

You can make the projected image larger by increasing the distance between the two pieces, or smaller by decreasing the distance.

5: Cameras

There are many different kinds of cameras, each with their own strength and weakness. In the following sections I will cover many of them which hopefully will give you some ideas of what you want to use.

We will look at how some cameras attach to some telescopes as well in case you have, or plan on purchasing one or the other.

5.1: DSLR (Digital Single Lens Reflex)

A DSLR is a Digital Single Lens Reflex camera, one that looks like what the professionals use. You look through a viewfinder in the back (the pentaprism) and see through the lens exactly what the camera sees. The lenses are removable and can be changed out to provide for different focal lengths (focal length is what determines how large an object appears to the camera) and capabilities.

Live view is the ability of the camera to show on the back screen instead of through the viewfinder what the camera sees out the front of the lens in real time. Not only does this make it easier to see brighter objects such as the moon, Saturn and Jupiter, but it also helps us focus (more on that later). It also works wonders on solar eclipses when you have a good filter installed.

An example of live view is shown with the moon in the following image.

Full manual control refers to the camera's ability to allow you to set all the options such as shutter speed and ISO without regard to what the camera thinks it should be. For most DSLRs manual mode is achieved by placing the top rotating dial to the "M" position as shown in the next image.

In the above image you may also note that there is a second lower dial. This is used to control the speed at which the camera takes images. The S stands for Single, or pressing the shutter release causes one image to be taken. The CL stands for Continuous Low speed which means that one press of the shutter will take an endless succession of images at low speed. CH stands for Continuous High speed. You should always use single in astrophotography.

Virtually all DSLRs also have the ability to be controlled remotely, either through software connected to a computer, through an intervalometer which is a device for taking a series of images automatically with no intervention from the user once the process is started, or by using a remote control.

Shown in the above image is an intervalometer for select Nikon cameras which costs less than $40 from many online vendors. It is also available for many other makes and models of cameras, and of course you can spend more, or less, on one depending on features and quality. This one will allow me to take hundreds of images over many hours.

You can of course also fire the shutter using software if you have a computer and the correct cables. The advantage of the intervalometer is that it is cheap (compared to a computer anyway), easily portable and has low power requirement (mine has shot more than twenty hours of images on the AAA batteries that came with it), and it fits into your pocket.

Many cameras can also be controlled using a wireless remote control, or even through a smartphone app.

Lastly, since DSLRs have removable lenses, they can be used in what is called "prime focus" as shown in the previous image which means a telescope can be used just as if it was a large lens. This is the preferred method because it has less distortion caused by objects in the light path (such as eyepieces which were never designed to be used to shoot images through). This method can also be used to mount a light pollution filter on the prime focus adapter assuming you purchase an adapter that is threaded for this purpose.

If you are even remotely really interested in astrophotography, even a cheap old used DSLR will open up a lot of possibilities to you. Something like a Nikon D70s (not the D70) can be had used for around $100 and will be a huge improvement over a phone, inexpensive point & shoot, or any other camera that cannot give you full manual control and be triggered with a remote. Maybe even splurge for something like the D80 which will give you live view as well.

The full manual control is especially important with lunar eclipses as the exposure changes dramatically during the course of the eclipse.

5.2: MLC (Mirrorless Cameras)

Mirrorless cameras (MLCs) are a relatively new addition to the camera stable and include some great features.

These cameras have interchangeable lenses, usually have full manual control, can have similar sized (if not identical) sensors to DSLRs and can produce almost, or as good images as DSLRs for most uses. Since MLCs have removable lenses they too can be used at prime focus as long as you can find a T-Ring (an adapter that bolts onto the camera as if it were that manufacturer's lens, and then screws onto a prime focus adapter).

One concern with MLCs is finding software or an intervalometer that will drive the camera so that you can remotely fire off a series of exposures.

The good news is that they all contain timers which can be used in place of an intervalometer or remote release.

5.3: Bridge cameras

The next type of camera "down" is the bridge camera which gets its name from the fact that it was designed to be in between, or a bridge between DSLRs and point & shoot cameras.

Bridge cameras very often have manual modes, but are different in that their lenses are usually not removable, which prohibits using the camera at prime focus although you can use it afocally (pointed through the eyepiece).

Bridge style cameras are my least favorite for astrophotography because they are almost as heavy as a MLC or DSLR yet since they cannot be used at prime focus, and most afocal adapters are designed for lightweight point & shoot cameras, that weight can cause some serious problems with image stability.

You can also have severe difficulty in finding software or an intervalometer to drive the camera. In fact, you may not be able to find these items or even a remote for a bridge camera.

If you have one of these and want to use it, make sure you know how to access the manual controls (for a lunar eclipse especially), that the solar filter is secured exceptionally well, and it will focus on the moon or sun when you need it to.

I have had issues with bridge cameras and point & shoot cameras that would not focus correctly once a solar filter was installed. Test it before the big day to make sure that it will focus correctly.

5.4: Point & Shoot cameras

Point & shoot cameras are the typical digital cameras in use today, for those who have a separate camera:

While point & shoot cameras suffer from some of the same issues as bridge cameras, they are light weight enough to be used with inexpensive adapters pointed at the eyepiece of a telescope. I have seen some remarkable images captured with little point & shoot cameras.

Many manufacturers produce different ways to mount a point & shoot to a telescope; my favorite so far is the Zhumell as shown in the next image. While a little large, it has the advantage of being so flexible I cannot imagine a small camera that could not be used with this mount. It is a few dollars more expensive than some adapters, but I would prefer to buy this item once rather than buy another and have it not fit a different camera I wanted to use down the road.

If you look closely at the above image you can see the knob at the top of the adapter which tightens a clamp onto the eyepiece (it must be a 1.25" eyepiece, 2" eyepieces will not fit) and a knob on the bottom of the camera which not only affixes the camera to the platform, but allows it to move forward and backward on the platform. Look closer at the bottom and you may see the small knob that allows the camera platform to move up and down. Not visible is the knob on the far right side that moves the platform left and right.

Make sure that your camera can disable the flash if you are taking pictures of a lunar eclipse, and if it cannot, that you find a way to completely cover the flash so that no light leaks out (a word of warning; flashes can generate a lot of heat, make sure you do not trap in the heat as well as the light or you could damage your camera and/or get burned). The flash can ruin your images and really upset other people if you are at the dark site.

Most point & shoot cameras have a tripod thread in the bottom so they can be mounted to a telescope and/or telescope mount just like a DSLR can using those threads. This allows the camera to track the sun (or moon) without having to point through a telescope lens. If you use this method with a solar eclipse make sure that the solar film you cover the lens with is secure.

Most of these also have a self-timer built in which you absolutely need to use to minimize vibrations and get sharper images.

5.5: Phone / Tablet cameras

These days almost everyone has a cell phone that can take pretty good images, can you just use that? Of course you can! There are some really neat mounts for it such as the iPhone mount by Orion shown in the next image. Originally for smaller iPhone, a little tweaking to the retainer clip is all that is needed to use it with an larger one as shown. Be sure the adapter you purchase will work with your phone.

With this adapter there are two knobs on the rear of the mount and one on the eyepiece clamp. This allows you to position the camera lens in the center of the eyepiece regardless of the eyepiece being used. I have found that if you use a light touch, you can actually tap the screen to take the picture without causing too much vibration, it works surprisingly well, although I recommend using a timer instead.

You can also use your phone or tablet to image an eclipse directly. A Lunar eclipse is easy enough and works best with a device such as the iPhone 7 Plus that has two cameras, one of which has a more telephoto lens than the other.

Find a way to mount the camera on a tripod or other device to make it as steady as possible as later in the lunar eclipse the images will be dark so the camera will expose the image for a longer period of time, making it more susceptible to vibration and shaking causing a blurry image.

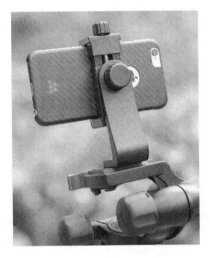

The phone holder shown above uses a standard ¼"x20 female threaded socket and screws instead of springs which makes the phone more secure. Twist one knob to loosen/tighten the phone in the mount, loosen the other knob to be able to rotate the camera and then tighten it back when done. It attaches to any tripod with a male ¼"x20 stud, which is virtually any ever made.

For a solar eclipse I still recommend a phone or tablet with two cameras, or with the highest resolution possible so you can crop the image. Then you need to outfit the camera for solar use.

5.6: Film cameras

Film cameras can still be used and will work just like DSLRs with the exception of having to shoot film. This makes stacking images more difficult (you certainly can take multiple images, scan them all in and stack them), and severely limits you because you cannot see the results and make adjustments immediately. You should remember however that this was the only way astrophotography was done for a long time.

As with DSLRs you can mount them piggyback, eyepiece projection, or prime focus.

5.7: Video Cameras

Lastly we have video cameras. Primarily in budget astrophotography we use webcams (a computer video camera designed primarily for computer to computer video conferencing) or modified webcams to capture images of the moon and planets. The theory is that by shooting hundreds if not thousands of images quickly, we can take the best parts of each frame and combine them together in one image. For this we use software such as Registax (discussed in more detail later).

This has become extremely popular and there are tons of different cameras that have been modified one way or another to be used in astrophotography. In fact, check out the do-it-yourself section of my book *Getting Started: Budget Astrophotography* for one such modification.

If you are not interested in modifying a web cam, and want something that "just works", most of the major astronomy companies make video cameras just for this very thing such as Orion's Starshoot Solar System Color Imaging Camera (shown in the next image), Meade's Lunar Planetary Imager, and Celestron's NexImage Solar System Imager. These types of cameras range from $99 up to about $199 and typically come with software just for astrophotography.

Webcams are cheaper but can require time and effort to modify so they will fit in the telescope, and require you to find software yourself. This is not necessarily a bad thing as you may prefer aftermarket software and some of the webcams are just as good if not better than some of the telescope manufacturer's cameras.

Webcams and the previously mentioned dedicated video cameras can get up close and personal with the planets because of their small sensors which act like a very high magnification eyepiece. Since as we discussed they use only the best part of each frame they also work well with barlows (a device that can increase the magnification of a telescope) to further increase the magnification, to a point. Increase the magnification too much and your image will start to get blurry.

An example of what the Orion Starshoot Solar System Color Imaging Camera can do is shown in the next image. This is the result of approximately a thousand frames stacked in Registax. While there are far better images of Saturn out there, this gives you an idea of what you can expect these cameras to be capable of.

You can even attach a standard video camera to a telescope and create a movie of the entire eclipse. The above image shows a Sony HD video camera bolted to the side of my primary telescope assembly during the June 2012 Venus transit and this was the exact same setup I used during the 2012 Annular eclipse.

If you look really close you can see the viewfinder is open and that there is a piece of cardboard over the lens. This cardboard has a round hole cut in it with a piece of Baader solar film covering the hole.

You do not have to attach the video camera to the side of a telescope, you could also attach it to any other type of tracking mount, barn door tracker, or even run it yourself.

Here is an example of what a video camera image of an eclipse looks like:

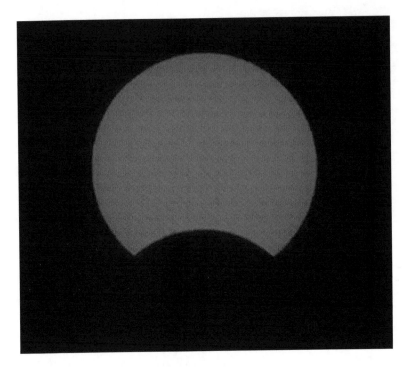

The entire video can be see here:

https://www.youtube.com/watch?v=gaaJkZVvYmI

Keep in mind that if you do not have a dedicated video camera you can use your phone, tablet, or any other device that shoots video, you will just have to find a way to mount it to your telescope, barn door tracker, or mount.

If you choose to run your phone or tablet in video mode and hand hold it, be sure you cover the camera with solar film securely and then come up with a way to prevent you from seeing the sun while you are recording. See the camera section on phones and tablets for information on how to do that.

6: Telescopes

Telescopes are absolutely not required to view an eclipse. If you already have one, or plan on purchasing one anyway, this section may help you decide what to buy or get the most out of what you already have.

Telescopes come in a wide array of types, sizes, features and prices. The first thing we need to make sure we remember is that while most people refer to the complete assembly as a telescope, we need to break it down into two parts, the telescope and the mount.

The telescope is the tube that light passes through and sits on top of the mount. The mount usually includes the tripod and any computer guiding that may come with it. There are telescope assemblies where the telescope and mount are not designed to be used separately but these are fortunately fairly few.

You do not have to have a telescope to dabble in astrophotography. There are many targets you can get wonderful images from with a 200mm or more lens. You could also play with wide field astrophotography which is a term used to describe taking images of large sections of the sky with any camera lens you may have already, or any you care to use.

If you want to do budget astrophotography then you probably will not be buying a telescope just for astrophotography, it will be used for visual with astrophotography being an occasional use. If this is the case then you can use pretty much any refractor quite easily, any Newtonian classified as an "astrograph" (meaning it is meant to be used taking pictures), or most SCT/MCT designs by having both a visual and imaging "back" (the part that holds the eyepiece or camera, respectively).

Dobsonians are not a good choice even for occasional astrophotography use as they rarely will focus with a DSLR camera attached in prime focus mode. More on that later.

If you want to get a little more serious but not break the bank, consider a small 80mm wide field refractor, something in the 400-500mm range. These can be had very inexpensively and adapt to many different kinds of cameras very easily.

6.1: Refractors

The first type of telescope is the refractor as shown in the next couple of images.

On the left of the above image set is an inexpensive iOptron 80mm refractor on a small go-to Alt-Az mount which costs around $300 delivered. On the right is a 110mm refractor with an 80mm refractor on top of it, all sitting on a fairly substantial go-to EQ mount which costs somewhere around $3000.

Refractors can be good for astrophotography (AP from here on out) work because they have no central obstruction, do not suffer from coma (an optical aberration in reflectors), do not usually need to be collimated (have the mirrors/objective lenses aligned), require virtually no cool down time, offer low wind resistance, have a higher Strehl ratio by nature (practical optical quality in the real world as opposed to theoretical), and a smaller area for their mass (this makes it easier for the mount to drive them).

The down side for a refractor is that inch for inch, they are the most expensive type of telescope. They also generally have a shorter focal length than SCTs (Schmidt Cassegrain Telescope) and longer than imaging Newtonians, which is neither good nor bad, but is a consideration.

Refractor Telescope

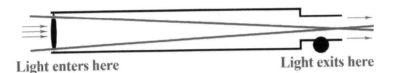

Light enters here   Light exits here

Note that the light just passes straight through; this maximizes the amount of light gathered per millimeter of aperture and minimizes problems inherent with bouncing light all over the place as mentioned below in the section on reflector telescopes.

Refractors are excellent beginning telescopes for astrophotography and small starter versions can be had for very little money. If you decide to get more serious about the hobby, your inexpensive first refractor can follow along as a guide scope for a larger refractor later on down the road making it a fantastic investment.

If you want to use this to image solar eclipses, simply measure the outside diameter of the front of the scope and buy a filter that slips over that end.

6.2: Newtonians

Reflectors can be good for AP work because they can easily offer longer focal lengths or faster focal ratios and are less expensive per inch. Some can also be much more compact than a refractor. Note that in the Newtonian design, which is the most common reflector type, the light has to bounce off of two mirrors that must be precisely aligned. The process of alignment is called "collimation", and needs to be done frequently. Also, the end of the tube through which light enters the Newtonian design is open to the air, which allows in dust, dirt, spiders, dew, and other things we don't want to talk about.

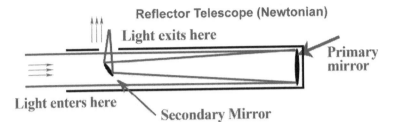

One common problem with most inexpensive Newtonians is that if you attempt to attach a DSLR at prime focus, you will not be able to bring the camera to focus. It will get close to focus when the focuser is all the way in, but it just does not seem to go in far enough. The solution here is to use a barlow (a device that plugs in where the eyepiece goes, and then an eyepiece plugs into it. This increases the magnification of the image) or replace the focuser with a low profile version. You could in theory also move the primary mirror forward but this is something that can be tricky so I do not really recommend it.

You can also image with one of these afocally by shooting through the eyepiece.

While an astrograph Newtonian (a Newtonian specifically made to image with) can be an excellent imaging solution I have found that it is rarely a good choice for someone just starting out on a budget. You certainly can get some good images if you are willing to work with it, and if you already have one it can be worth it.

Most solar filters for these do not cover the entire opening but instead have a black cover that goes over the front with a much smaller diameter hole in it which contains a solar filter.

6.3: Dobsonians

Dobsonians are a type of Newtonian telescope whose rear end sits on a special Dobsonian mount on the ground. These have the advantage of being very inexpensive per inch of aperture, but most of the ones you purchase commercially are not tracking at all which prohibits log exposure astrophotography but is fine for eclipses.

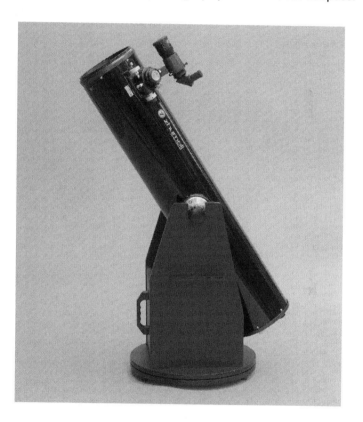

One of the largest issues is that in order to get a DSLR (or any camera at prime focus) to work, you will find that you cannot get the camera to come to focus. This is because the sensor for the camera is farther back than an eyepiece would be so you cannot move the camera in far enough to achieve focus.

The fix is to use a barlow but that will add to the focal length of the telescope and make the image correspondingly darker and harder to image. This is not much of a problem with the moon, sun and using video to image planets and so works fine for eclipses. It can however make any deep space targets all but impossible.

Most solar filters for these do not cover the entire opening but instead have a black cover that goes over the front with a much smaller diameter hole in it which contains a solar filter.

6.4: Schmidt-Cassegrain & Maksutov-Cassegrain

Some scopes are hybrid scopes such as Maksutov-Cassegrains (MCT) and Schmidt-Cassegrains (SCT) which combine some of the qualities of both a refractor (being sealed and sometimes having lenses) and a reflector (multiple mirrors). These typically have the advantage of being much smaller than Newtonians, and are sealed against dust. They have the disadvantages of being slow scopes (higher focal ratios, longer focal lengths), having to be more frequently collimated than refractors and having long cool down times (longer than Newtonians since they are sealed).

One problem with MCT telescopes is they tend to have very long focal lengths which make the images they produce rather dim, presenting more challenges to astrophotography. A solution for this is to attach a focal reducer to the scope which reduces the focal length and brightens things up (this works well for SCTs too). An issue here is the MCT accessories can be rather expensive and hard to come by. A solution to this might be to purchase a Baader SCT/T2 Universal Adapter #2958500 for about $50 which allows you to use less expensive and more readily available SCT accessories on your MCT. While it fits a lot of MCTs, be sure to check and make sure yours is one of them before purchasing this adapter.

If you plan on doing planetary astrophotography (taking pictures of the planets), then the long focal length of an SCT or especially an MCT may actually be a nice advantage as the planets are relatively bright anyway.

If on the other hand you are just concerned with the eclipse, then the long focal length may actually be a serious problem that requires the use of a focal reducer.

The next question is always, which one is best, an SCT or MCT? That is much like asking what kind of vehicle is best. For some people, a pickup is better because they are constantly hauling things. Others may need a minivan to put all the kids in. Still others may need a small, light car for fuel economy.

The old adage of "bigger is always better" or "aperture is king" in visual astronomy is only part of the story in visual astronomy and is even less applicable in astrophotography. My 4" refractor is regularly imaging the same targets as people who use an 8" SCT, and doing just as well or better. The key is getting a scope with the right combination of features that works well for you.

With a budget in mind, I would put a nice 80mm refractor at the top of the heap for beginning budget astrophotography, keeping in mind it has serious limitations with visual astronomy.

If your budget is a little larger, spend that money on a good mount and a high quality light pollution filter as the telescope is probably the last thing you should spend more money on.

If however you already have a scope like this then it will do fine for imaging eclipses although make sure that the entire sun will fit into the field of view, particularly with MCTs.

Most solar filters for these do not cover the entire opening but instead have a black cover that goes over the front with a much smaller diameter hole in it which contains a solar filter.

7: Motorized mounts

The idea behind a motorized mount is twofold: to provide a stable platform for viewing or imaging, and to track celestial objects to keep them in the eyepiece or camera without your intervention.

Some even have capabilities such as go-to where you tell the telescope what object you want to look at (usually by selecting it from a list on a handheld controller) and it automatically points at the object for you.

Slightly less advanced (and less expensive) is the push-to controller which works just like the go-to except you act as the motors that move the telescope. There is usually a system of arrows or lights to indicate which direction you should push the telescope to get it to aim at the object you want to see. Once the telescope is pointed at the object, the indicator will let you know.

Lastly is the pure tracking mount. These do not know where they are pointed, have no idea where anything is, and could not care less. What they do is have motors that move them in such a way as to keep an object in the viewfinder or camera once you get it there to start with.

Tracking mounts are the least expensive and have been around the longest but current models are also less desirable today just from a stability point of view.

Motorized mounts are in no way required for viewing or imaging an eclipse, but they can make it easier.

Let's take a look at some mounts.

7.1: Altitude Azimuth

The next type of mount is the Alt-Az mount. Most inexpensive telescope mounts are of this variety. They move up, down, left and right. They can follow objects in the sky just fine, but suffer from field rotation.

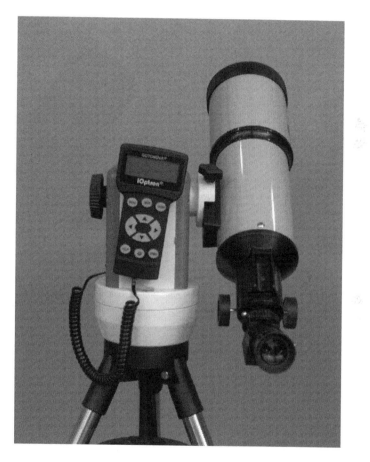

In addition, since these were designed for visual astronomy and not astrophotography, they do not usually have nice smooth tracking motions but instead seem to be a little jerky in their motions. This is usually slight enough that you cannot see it with the naked eye, and should not present a problem with an eclipse.

One thing you should know is that some of the more expensive Alt Az mounts can use what is called a "wedge" as shown in the next image. This can make an Alt Az mount behave like an EQ mount (discussed in a moment). The telescope computer must be able to handle "EQ mode" (using the telescope like an EQ).

Wedges, and EQ mounts as we will discuss later, are primarily for long exposure astrophotography and have no bearing on eclipses.

The wedge picture above is mounted to what is called a pier, basically a pole mounted into the ground to increase stability. This particular one is a steel pole bolted into the concrete which is more than sufficient for visual or short exposure astrophotography work.

7.2: Equatorial

The next type of mount is the German Equatorial mount, sometimes called the GEM, sometimes called an EQ. Recently the Japanese manufacturers have released different designs that accomplish the same tasks as the GEM but with a wildly different design. So instead of having GEMs and JEMs, I just call them all EQ mounts.

EQ mounts will work exceptionally well for eclipses but they are also massive overkill for just that use. If however you are interested in astrophotography beyond just eclipses, they will serve you well.

EQ mounts are designed to track the motion of the stars including field rotation. This means, in theory, you could take a single photo over several hours with no star trails (have perfectly round stars). I say in theory because this is impractical for a host of reasons that are far too in depth to be discussed here.

These mounts can start at around $500 or so for just the mount, no telescope, no adapters, nothing. If you are serious about astrophotography you will most likely own one or more of these mounts. My favorite inexpensive EQ mount right now is the Orion SkyView Pro go-to at about $850. While this is hardly a low budget mount for most people, it certainly is for midrange astrophotography and I include its information here just as a point of reference.

These mounts are lined up pointing towards the celestial pole, a process called polar alignment, as shown by the arrow in the following image. This is what allows them to follow the arc of the objects through the sky.

Another interesting feature of EQ mounts is that they use weights to balance the scope assembly to make tracking a target smooth and easy. If you look at the previous image you will see the two round black weights at the end of a silver shaft extending from the bottom of the mount to the left of the tripod.

EQ mounts are rated in how much weight they can carry, excluding the counterweights. What this means is how much weight can go on the top as shown in the picture which includes things like the telescope, guide scope, cameras, dew prevention devices, etc. This is balanced out by roughly equal weight on the bottom using the weights.

One disadvantage of EQ mounts is that they cannot shoot all night uninterrupted as this would cross the meridian (an imaginary line in the sky directly overhead). They can shoot up to the meridian, and then we preform what is called a meridian flip where the scope flips directions while pointing at the same place, and then it can continue on shooting for the rest of the night.

One other form of EQ mount is the one built only for cameras like this SkyWatcher Adventurer:

These mounts attach to a tripod and then have a screw that screws into the tripod socket on your camera. Once adjusted correctly they can allow you to take images which track an object in the sky.

This kind of mount is cheaper and far more portable than a standard type of mount however they have a serious weight limit of approximately ten to eleven pounds. This means if you have a large heavy lens, you may not be able to use this type of mount.

For a standard DSLR and 300mm inexpensive lens, or any smaller camera type such as point & shoot, bridge, or MLC, this would be an excellent item although overkill for just eclipses.

7.3: Tracking mounts

There is a difference between computerized mounts and tracking mounts. Tracking scopes have no idea where anything is, they simply track whatever target you point them towards. A major disadvantage to newer inexpensive tracking scopes is that they are usually cheap. I don't mean just inexpensive, I mean cheap. Cheap as in you will get very frustrated with them and want to do bad things to them. It is amazing what an axe can do...... oops, forget I said that.

Back many years ago only high end telescope mounts could really be fitted with tracking motors, and both the mount and motor were anything but cheap, or inexpensive. Today that has drastically changed and many less expensive mounts either already have tracking capability, or can be fitted with it. Unfortunately these cheap tracking mounts will not work for astrophotography in most cases.

One cheap tracking mount I have had the displeasure of dealing with was so bad that I could take better images on a standard tripod with no tracking of any kind than I could on it.

There are however nice tracking mounts out there such as the Vixen Polarie, iOptron SkyTracker, Astrotrac and SkyWatcher Adventurer in the previous section. These can be very accurate and quite robust; they also start at about $400 without a tripod.

Do not take this as meaning that computerized telescopes are perfect tracking devices. Cheaper ones can be almost as bad as the cheap tracking mounts.

8: Making things you can use

There are a lot of things you can make at home instead of buying for taking pictures of an eclipse. In fact, I could probably write a book just for that, but let's not.

What follows is a few items from my other book *Getting Started: Budget Astrophotography* that I thought would be useful specifically for eclipses.

If you wind up getting interested in astrophotography as a hobby, you might want to pick up my other books for more projects like these.

8.1: Glass solar filter for a camera lens

During the annular solar eclipse of 2012 I wanted to use a DSLR with a 300mm lens as a secondary imaging setup so I could capture as many different shots of the eclipse as possible. My problem was I needed a 55mm solar filter and could not find one, so I decided to make one. This project originally appeared in my book: *Getting Started: Budget Astrophotography*.

Let me start by saying any time you point anything at the sun, you need to be extremely careful. What I am about to describe worked for me, but I make no representations about its safety. Trying this idea yourself could result in damage to your camera, temporary or permanent blindness, or worse. I take absolutely no responsibility for anything that happens. You have been warned.

While some may disagree, I am not a complete idiot, so I ordered some Baader solar film and tried to figure out how to mount it to the front of the camera. What I found was for about $5 I could order a clear UV filter online (with free shipping I might add) and then I could attach the film to the filter.

I used a piece of paper to trace around the smaller threaded side of the filter and then with scissors I cut a slightly smaller circle. Then after making sure that paper circle fit correctly inside the filter ring I cut the solar film to size.

Now that I had the solar film the right size I had to attach it to the glass and I used a tiny amount of a paste glue in a ring around the very outside edge of the glass and let that dry.

Once this was complete I still had one problem, the edges of the film were not perfectly against the side making a seal, so some sunlight could get past the edge of the film, this was unacceptable. To solve this, I found some black paint and made a paint ring around both sides of the glass, using multiple coats. This created a nice barrier about 1/8" around both sides for safety.

As you can see from the previous image I put the film on the inside of the filter so wind, sand, etc would not touch it. The last thing I needed was the wind to get under an edge and rip all or part of the film off.

I checked my work with the brightest flashlight I had available.

In the field to make sure the filter worked, I first held the filter between the sun and a piece of white paper to look for bright spots, then I held the filter up to the sun wearing quite dark sunglasses and checked again. Once fitted to my camera I used the camera's live view feature to put the sun in the center of the field of view.

Under no circumstances should you look through a camera pointed at the sun, the magnification of sunlight provided by the camera could cause permanent blindness or worse. In addition, you should never look directly at the sun without proper and well tested equipment.

So how did it work?

I would say pretty well! Since the sun was reasonably small in the field of view anyway so I had to crop away the edges which is where any problem would be from the glue and black paint, the center remained quite sharp.

8.2: Cheap and easy solar shade hutch

When doing solar photography and using a laptop you need this so you can see your laptop screen in the daytime. Without it, you will have an extremely hard time seeing what is happening on your laptop screen. This project originally appeared in my book: *Getting Started: Budget Astrophotography*.

The solution to the problem is simple, build a little hutch to keep the light off the items on your table. We have all used the discarded shipping box for this, and that does indeed work, once or twice. Here is a better way.

It seems every years there is a plethora of yard signs everywhere you look. They contain advertisements for things for sale, political ads, etc. Most of these have one thing in common; they are made from corrugated plastic instead of cardboard. The advantage here is that the corrugated plastic is much more robust than cardboard, and is immune to moisture such as dew. With proper care there is no reason a hutch like this could not last many years before being replaced.

I plan on using mine for both dew prevention and keeping out light while doing solar work so I decided to paint mine flat black using a can of spray paint I bought for about $4. Be sure to use flat black paint because it does not reflect light like regular gloss black paint does.

The stands they are already on are great for painting, just stick them in the ground and start spraying!

Once the paint looks good (don't be too concerned with looks) remove the stands by just pulling them out. Now measure how wide and deep you want your hutch to be and start measuring and marking the signs to make the top, rear and two sides you need for the hutch.

Once the pieces are cut and fitted together, I used loose zip ties run through drilled holes as the hinges.

Note the notch out of the corner of the side piece, this allows clearance of the top zip tie. Also note how loose the zip ties are, this allows the thing to fold up really easy. You need the top attached to the back by at least three zip ties, then each side is attached only to the rear with two zip ties each. The sides should not be attached to the top.

If you need to run cables to the laptop, simply cut notches out of the bottom of the back or sides to run the cables out.

This completely covers my 17" laptop and yet packs up very small.

Folded up the entire thing is 21"x12"x1/2" and weighs almost nothing. What's more, it is durable, light proof, dew proof, easy to set up and easy to pack. This awesome little hutch can be yours for less than $5 including paint and about an hour of work.

8.3: 36 Amp hour field power pack

One thing you may need out in the field is power. You might get lucky and have access to all the AC power you could need, or you may need to run your setup off of batteries. If you run off of batteries you need to run your scope, mount, and accessories.

I decided that even though cheap "jump start" battery packs are available, I would be willing to bet I could do it just as well or better, and cheaper. What I set out to do is run my telescope and dew heater for as long as possible. It also needed to be easy to work on should anything go wrong (batteries do eventually die). Here is what I came up with.

The above image shows an $8 toolbox from Home Depot, four 9Ah 12V batteries (about $20 each), 1 roll of 12 gauge power wire (I actually got two, a red and a black, but forgot to put the black in the picture, about $3 each), a package of crimp on female spade connectors (about $3), and a 2 port 12v cigarette lighter socket (about $6).

The first issue is to make sure that the toolbox you buy fits all four batteries well.

Note I have some space in between the left side and right. I will put a piece of wood in between the sides to keep everything in place.

Next is fitting the power panel to the lid of the box. I ran into my first problem here in that there was no one place I could put it, so I had to decide where I could mount it and not really hurt the structural integrity of the lid. I need the lid to remain as strong as possible since I will be lifting four batteries by it.

I finally decided I had to remove one of the supports in the lid so I carefully cut the ends lose and then grabbed the plastic with square nosed pliers and twisted. This did a fairly good job of removing the plastic as shown in the next image.

Now I needed to smooth the edges where I removed the plastic so the power panel would fit in right up against the support on the right.

I used a rotary tool with a small round sanding drum to remove the excess plastic and smooth everything out.

Next I marked and drilled the holes for the two power ports. I used an Irwin Speedbore in 1" to drill the two holes and then used the same small Dremel sanding drum to enlarge them very slightly.

Now I needed a way to secure the ports into the lid and another problem popped up. The plastic was too thin for screws. I thought about bolts but that required rather large holes (another problem) or very small bolts (which needed larger washers, another problem with clearances). Then I thought of zip ties.

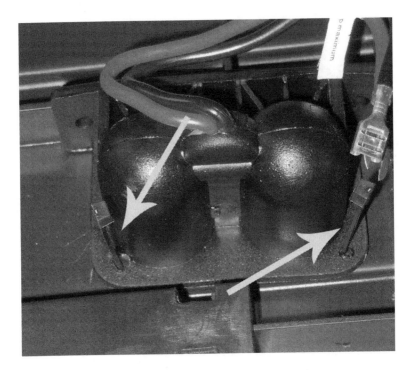

I drilled little holes, just big enough for the zip tie, and then ran the zip tie through the bottom, using a second zip tie to secure it from the top as shown in the next image.

I then did that two more times to the opposite side so that there are a total of four zip ties holding the assembly to the top of the toolbox.

Now it was time to start wiring everything, so I stripped the ends of the wires coming from the power panel. Then I cut and stripped four red and four black wire lengths. Finally I soldered the whole thing together as shown in the next figure.

Note that I placed some heat shrink on the wires before soldering them, the heat shrink is in the upper right of the picture, those lighter gray sleeves over the wires.

Once the soldering was done I slid the heat shrink tubing down over the connections and used a lighter to shrink them. I then added a zip tie to the bundle of four wires as a strain relief. This is so the wires would not try to spread apart and rip the heat shrink.

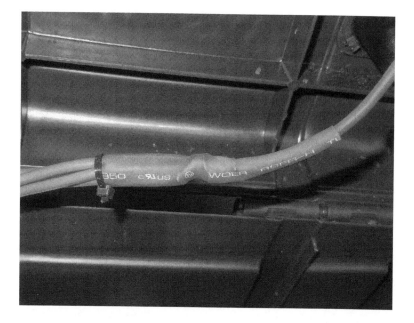

Now all I had to do was strip off a little of the other end of each wire, crimp the ends on them, and attach them to the batteries.

One big advantage to this project is that if I needed to start out smaller/cheaper, I could build it exactly the same way but only install one battery. Then as I needed more power and had more money I could buy another battery and just plug it right in.

Since the batteries are easy to remove, I plan on removing them and charging them with my 12v automotive battery charger (it has a low setting just for batteries like these). I could also get a 1A, 13V DC (check the actual power output using a meter, it could put out 15V and be marked 12V, you need something that puts out 13-13.5V DC while charging) power adapter that plugs into an AC outlet. Then put a male 12V cigarette lighter end on it. This could plug directly into the wall, and then plug into the female port on my box.

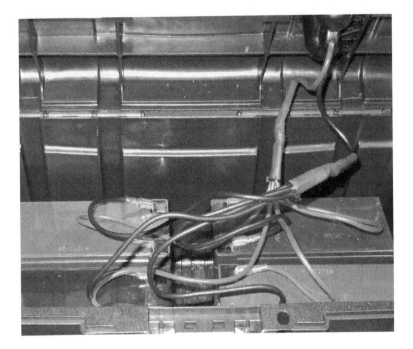

36Ah is a lot of power. How much will it run? My pretty hefty Orion Sirius EQ mount uses about 2A, my DIY dew heater less than 1A, so let's say 3A total. This battery pack should easily run that setup for over ten hours on a single charge on a warm night. Cold weather reduces the available battery power but even then, you should be able to get a full eight hour night off a single charge without a problem.

9: More information

More information on eclipses can always be found at NASA's websites:

https://eclipse.gsfc.nasa.gov/

Tons more information on astrophotography as well as links to my books on the subject and YouTube channel can be found on my website at:

http://www.allans-stuff.com

Solar filters and other equipment can be purchased from:

http://thousandoaksoptical.com/

http://agenaastro.com/

Following this page is an index, glossary, a section on other books I have written and a couple pages for you to write notes.

The glossary here is slightly more extensive than this book requires but I left in terms you will not find in this book in case you decide to get more into astrophotography, or talk to an astrophotographer about taking eclipse pictures. I hope they help.

9.1: Index

9.2: Glossary

A/D converter (ADC) - Analog to digital converter. A camera sensor records light as an analog signal which the A/D converter then converts into digital information.

Achromat – A type of refractor typically with two lens elements to correct for chromatic aberrations. This type of scope is not well suited for astrophotography.

Afocal - A means of taking an image through an eyepiece of a telescope without removing the lens from the camera.

Alt/Az - Altitude Azimuth, a type of telescope mount that moves up and down, left and right as opposed to the smooth rolling motion of an EQ mount which accurately tracks the motion of the stars around the earth.

Amp glow – Amp glow is the glow that some cameras show on a long exposure image. This usually manifests itself in the corners of the image first and then can spread towards the center. A moderate amount of this can be removed using dark frames. Severe cases cannot be corrected.

Aperture - In telescopes, the diameter of the opening at the front of a telescope, usually measured in millimeters. Can also be measured in inches for larger scopes. In camera lenses there is a diaphragm inside the lens that controls the aperture which is sometimes referred to as an F-Stop.

Apochromatic (APO) – A type of refractor extremely well adjusted to remove most or all chromatic aberrations which makes it excellent for astrophotography uses. Can have two, three, or more lens elements. Higher end versions almost always have three or more elements.

Arc Minute – There are 360 degrees in the sky as it goes 360 degrees around us. One arc minute is $1/60^{th}$ of a degree.

Arc Second – Is equal to $1/60^{th}$ of an arc minute.

Artifacts - Errors or unwanted signals in the image.

ASCOM - abbreviation for AStronomy Common Object Model and is a standard in the astronomy equipment industry for control interface design of astronomical equipment such as mounts, focusers, motorized domes, etc.

Astrograph - A type of Newtonian telescope that is designed specifically for astrophotography.

Astrometry – Extremely precise measuring of objects like comets and asteroids.

Astrophotography - Photography of objects in the sky.

Autoguider - A camera and associated equipment used to increase the accuracy of the mount in tracking the stars.

Audio Video Interleave (AVI) – A wrapper for computer video files, can contain a variety of different formats, typically video for Windows formats, and has a file extension of .AVI.

Back Focus – The necessary distance needed to be able to attach a camera onto a telescope focuser, and be able to bring the image projected onto that camera's sensor into focus.

Backlash – Unwanted spacing between gear assemblies usually resulting in some "play" or "slop" with the device. This is normally used to describe issues with a mount but can be applied to anything with gears.

Baffles – Ridges running around the inside of the light path in a telescope to prevent the scatter of light inside the telescope and provide an image with greater contrast.

Bahtinov mask - A mask or cover that goes in front of a telescope with a specific pattern of slits designed to provide easy focusing of point light sources such as stars.

Barlow - An optical device that increases the magnification or reduces the field of view, depending on how you look at it. This trades some image quality and light for more magnification. These plug into the optical train just before the eyepiece.

Bayer matrix - In color one shot cameras (any camera that produces a single color image in one exposure) the pixels are grouped in groups of four, one red, one blue and two green. These are combined to generate the color information for that area of the image. The Bayer matrix is the array of colored filters over the pixels that accomplishes this.

BFA – Bayer Filter Array, see Bayer matrix above.

Bias frame - An image taken with the highest shutter speed possible on a given camera at the same ISO and temperature of the light frames. This is used to subtract the camera's electrical signal present in every frame it takes from the final image.

Binning – A process of combining multiple pixels in order to boost sensor sensitivity at the expense of resolution. For example, 1x1 binning means each pixel counts as one pixel and is in effect not binned, 2x2 binning would take a square of 4 pixels and combine them into one "super pixel".

Binos - Short for binoculars.

Bino-Viewer - A device that allows attaching two eyepieces to a standard telescope so you may view objects in stereo.

Bit - A single bit can be either on or off, representing either 0 or 1. Computers use this as the basic language of everything they do.

Bit depth - This describes a measurement of something like the number of colors an image can contain and is base two mathematics. An example is a 1 bit scale will contain two possible combinations, a 2 bit scale will contain 4, a 4 bit scale will contain 16 and an 8 bit scale will contain 256 bits.

Black point - An area of an image that represents absolute black.

Blooming - In a camera, once a pixel has received as much light as it can handle, the voltage can spill over into adjacent pixels causing them to be brighter than they should.

Bortle scale – Astronomer John Bortle developed a scale of nine levels which represents the "true darkness" of a site, or the amount of light pollution present.

Bulb exposure – A bulb exposure is an exposure where as long as the shutter button is held the camera continues the exposure. DSLRs and other cameras can be used in this mode.

CCD - Short for Charged-Coupled Device, a type of sensor used in digital cameras. In astrophotography it is usually used as a reference to a camera designed and used specifically for astrophotography as opposed to a digital SLR or other multi use digital camera.

Celestial equator - An imaginary line which is basically the equator of the earth projected up into the sky.

Center mark – A dot placed exactly in the center of the primary mirror of a Newtonian to aid in collimation.

Chromatic aberration – Chromatic aberration is the "glowing" or "fringing" of light around bright objects in a telescope. This is caused when light passes through the optical path it is split into its component colors and then rejoined imperfectly at the focal point.

Clip - Clipping an image means you have cut off one end or the other of the image's ability to record data (as can be shown in a histogram). Clipping the highlights for example means that area of the image is pure white and cannot contain any detail. Clipping the darks means that part of the image is pure black and contains no detail.

CMOS - Complimentary Metal Oxide Semiconductor. In astrophotography, a type of sensor in a camera.

Collimation - The act of aligning the optical components of a telescope to make sure all parts of an image combine correctly into one sharp image.

Coma - An optical defect normally present in reflector telescopes that can cause point light sources such as stars to appear to be out of round, presenting like they have the tail of a comet.

Coma corrector - An optical device for reflector telescopes to correct for coma aberrations.

Convolution – A mathematical method of multiplying arrays of numbers to get a third array of numbers. Used in image processing to stretch or resize images.

Corrector plate – The lens on the front of an SCT type telescope that corrects for the spherical aberration created by the spherical mirrors used in that design.

Counterweight – A weight, usually on an equatorial mount, used to balance the weight of the telescope and associated hardware.

Crayford focuser – A telescope focuser that uses smooth bearings and rollers as opposed to gears used in rack and pinion style. They usually come in dual speed (coarse and fine adjustments) and can have adjustable tension.

CRW/CR2 - Canon's RAW image format.

Dark frame - An image taken at the same ISO, shutter speed and temperature as the light frames but with the lens cap/scope cap on, or the shutter closed. This is used to detect the thermal signature of the camera's sensor at these setting so they can be subtracted from your final image.

Dead pixel - Opposite of a hot pixel, a pixel that is stuck in the off position and registers as black regardless of the amount of light applied.

Declination (DEC) – Celestial coordinate measured from the celestial equator north and south of that line, from +90 degrees to the north to -90 degrees to the south, zero being the celestial equator.

Deconvolution - A method of image enhancement that corrects for the bad effects of convolution. This can substantially increase fine details in an image.

Dew heater - Usually a strip that heats up and is wrapped around a telescope near the optics. This warms the optics and prevents dew from forming.

Dew shield - A device attached to the end of a telescope and is like a hollow extension of the telescope tube. This delays the objective from collecting dew, and reduces the intake of extraneous light sources.

Diagonal - A device that has a mirror inside and reflects the image at a 45 degree or 90 degree angle for easier viewing. One side goes into the focuser, the other end holds an eyepiece.

Diffraction - As light passes through a telescope it passes through openings. As light gets near the edges of these openings it is diffracted. This causes stars to appear larger than they actually should.

Diffraction limited – Term used primarily by telescope manufacturers that says that the telescope should perform so that any defect seen will be with the physical characteristics of light and not optical problems with the telescope.

Dispersion – Cause of chromatic aberrations. Prism effect, when light is spread out into its spectrum from white light.

Dobsonian - a type of telescope mount, but usually used as a reference to the entire telescope assembly. These are usually larger Newtonians mounted onto a base that sits on the ground and moves as an alt/az. Like regular Newtonians these are not well suited to astrophotography due to not having enough backfocus.

Doublet – A refractor telescope with two objective lenses.

Dovetail - A metal rail that attaches to the bottom of the telescope, usually by rings that clamp into the telescope tubes or bolts into the bottom of the telescope, which can then be quickly and easily attached to the mount's clamp. Popular dovetail types include Vixen and Losmandy.

DSLR - Digital Single Lens Reflex camera. A type of camera where the user actually looks at the same image that will be recorded on the sensor by means of a mirror and prism that reflects the light from the lens through an eyepiece. When the shutter is opened to take the picture the mirror swings out of the way, the eyepiece goes black as it is no longer receiving the reflected image, and the sensor is exposed.

DSS – Short for Deep Sky Stacker, very popular free program generally used by beginning astrophotographers for stacking images.

Dynamic range - The range from brightest to darkest that a camera can record.

ED – Extra low Dispersion, optical glass corrected for chromatic aberration.

EQ/Equatorial Mount - A type of mount specifically designed to track the stars as they travel around the earth compensating perfectly for their arc in the sky.

Ephemeris – Detailed positional information about planets, their moons, comets and asteroids.

Eyepiece - An optical device that focuses the light exiting a telescope tube in such a way that you can view it with your eye. These typically contain many lens elements in a round cylinder that is inserted into the focuser. The eyepiece can be made to magnify or reduce the image size.

Eyepiece projection - A method of taking a photograph through the eyepiece of a telescope without a lens on your camera. This uses a specific adapter. This can come in handy on telescopes that cannot reach focus using a prime focus adapter.

F-Stop - When using a camera with its lens installed, the aperture is adjustable and is commonly referred to as the F-Stop.

Field flattener - An optical device used primarily on refractors to make sure that the image arrives at the camera sensor perfectly flat. This prevents elliptical images of stars in the corners of the images while the stars in the center may be perfectly round.

Field of view - Commonly represented as FOV. The area of the sky that you can see at one time. Longer focal lengths (more magnification) generally show smaller areas of the sky and hence a smaller field of view. Eyepieces with smaller numbers cause the same effect.

Field rotation – The effect of the image being blurred from the rotation of the sky. This can happen when you use an Alt/Az mount to take long exposures since the Alt/Az mount does not rotate the camera like an EQ mount does.

Filter – A filter is a piece of glass (or Mylar in some solar filters) that alters the light coming through the telescope before the eyepiece or camera. A filter is used for removing light pollution, enhancing certain colors, shooting color images with a monochrome camera and many other tasks.

Finder - A small telescope or other pointing device that helps you quickly orient your telescope towards a particular target. Similar to a gun sight.

Firmware - The software a device uses to tell it what to do. For example, your GoTo telescope software in the hand controller is called its firmware and can be updated on many devices.

FITS format - A file format designated by .FIT (such as .TIF, .GIF or .JPG) specifically designed for scientific purposes. Like RAW or TIF files this stores raw data that does not degrade from repeated editing as do formats such as .GIF or .JPG.

Flats/Flat frame - An image taken with even illumination over the front of the telescope and exposed to present a neutral gray image. This must be taken with the exact same setup as your light frames (same focus setting, same filters, etc) and is used to remove vignetting.

Focal length - The length of a line following where the light travels through a telescope, this is important for calculating parameters such as the FOV and magnification.

Focal plane – An inferred plane at the point where the image from the telescope comes to focus. A camera's sensor is mounted so that it is at the focal plane.

Focal ratio (FR) – The focal length divided by the aperture of the primary objective of the telescope.

Focal reducer - An optical device which reduces the effective focal length and increases the field of view of a telescope, seemingly reducing the magnification. This is usually mounted into the focuser before any eyepieces or cameras.

Focuser - A piece of equipment mounted on the telescope where the light exits. Eyepieces, diagonals, barlows and cameras are mounted into the focuser. Its job is to move the eyepiece/camera/etc back and forth until the light comes into focus at a specific point (your eye or the camera sensor).

FOV – See field of view.

Frames per second (FPS) – The number of image frames captured per second by the device, used in video capture devices.

Full well capacity - A measurement of the total amount of light a photosite can store before saturation occurs.

FWHM – Full Width Half Maximum. The measurement of the angular apparent size of a star, usually used to get the size as small as possible in an image which represents the best possible focus.

Gain - This is a multiplication of the incoming signal. For example, if one photon enters a camera and hits the sensor, setting the gain to 2x will cause the digital signal sent from the camera sensor to say that two photons hit the sensor. Increasing the ISO of a digital camera is increasing the gain.

German equatorial mount (GEM) – Another name for the equatorial mount.

GoTo – A telescope that when properly aligned can point to a celestial object automatically when selected from a catalog or menu.

GPS – Global positioning system, a device or feature used to determine your exact location on the planet.

Grayscale - An image recorded in black, white, and variations of gray with no color information.

Guiding - The act of following a star or other object using either manual corrections (as was the case back before GoTo and tracking mounts) or automatically using guiding equipment such as an autoguider.

Hand controller (HC) – The handheld device used to control your telescope's mount.

HDR - High Dynamic Range. You can use different exposures on different images and sandwich them together to show an image that has too much dynamic range to be captured in one single exposure. M42 is a prime example of a target that needs HDR processing: if you expose correctly for the faint dust lanes on the outer areas, the central core is blown out or clipped; if you expose for the central core, the outer dust lanes are clipped into blackness and can not be seen.

Highlights - Areas of maximum brightness in an image.

Histogram - A graph that shows how an image is exposed. In a normal grayscale histogram the left side is absolute black, the right side is absolute white and there is usually a hump in the graph display somewhere near the center showing the exposure of that image. Color works the same way but shows the intensity of the red, blue and green color channels.

Hot pixel - Opposite of a dead pixel. A pixel that shows exposure information even when shot in complete darkness.

Illuminated reticle eyepiece – An eyepiece with an illuminated crosshair or other centering marker used for precise centering of targets in the field of view.

ISO - International Standards Organization, used to measure the "speed" of film, or the sensitivity of a sensor in a digital camera. As ISO increases, less light is required to "expose" for a given image. This also reduces the signal to noise ratio, increases noise, and reduces the bit depth possible in the image.

JPG/JPEG - Joint Photographic Experts Group. A file format denoted by .JPG (such as .TIF or .GIF) that is very common in digital cameras. Using this format should be avoided because it uses a lossy compression format to reduce file size. This results in huge losses of information and makes it virtually impossible to process well for astronomical uses.

Light frame - A standard picture. Every regular picture you have taken with a regular camera of birthdays, friends and family are all what we call light frames. These are the frames you work with that contain your image data.

Light pollution – Stray light from street lights, signs, windows etc that shine or are reflected up into the air. This is scattered by contaminates and humidity in the air and create a glow effect around cities making it difficult to see outside the atmosphere.

Light year – The distance light travels in a year through a vacuum, approximately 5.87 trillion miles.

Limiting magnitude – The measurement of the dimmest star you can see at zenith which takes into consideration all parameters such as light pollution, weather conditions and optical devices used (if any).

Lossless compression - Certain file formats such as PSD and TIF employ compression methods that preserve 100% of the data while decreasing the file size.

Lossy compression - Formats such as .GIF and .JPG use lossy compression which throws away data that it does not think is needed to display the image.

LRGB - When shooting a monochrome camera and creating a color image you need to shoot at least one image with a red filter, one image with a green filter and one image with a blue filter. These are combined together into one color image. The L in LRGB stands for luminance and is used to increase detail in an image. The Luminance frame is the detail frame and can be shot in very high resolution. The color can be shot at lower resolutions and combined with the luminance to create a high resolution color image. You can use this idea to increase your ability to stretch images as well.

Luminance – The recording of brightness or intensity of light. Typically this is the high resolution/detailed portion of an image.

Magnitude - A measurement of the brightness of an object. An increase in one magnitude is approximately 2.5 times as bright. The lower the number on the scale, the higher the magnitude.

Maksutov Cassegrain telescope – See MCT below.

Maksutov Newtonian – Similar to a Maksutov Cassegrain except they are designed as a Newtonian configuration with the focuser near the front of the scope.

MCT - Maksutov Cassegrain Telescope, a type of telescope that has a sealed front end which is actually a corrector lens called a meniscus, two mirrors and has its eyepiece in the rear.

Megapixel - Roughly one million pixels.

Meridian - An imaginary line dividing the west and east halves of the sky running from the north celestial pole directly overhead to the south celestial pole.

Meridian flip - Meridian Flip is the act of re-orienting the scope on an EQ mount so it can continue to track past the meridian. This "flips" the scope around to pointing the other direction at roughly the same spot on the meridian. Going past the meridian without flipping can cause the scope to run into the mount, cables to come loose, and many other really bad things.

Micron – One millionth of a meter or 0.001mm.

Mirror cell – The frame that holds the primary mirror assembly.

Mirror lock(DSLR) – Some cameras have the ability to lock the mirror in the up position to minimize camera vibration when the shutter is tripped. This can be very useful shooting brighter objects like the moon but is ignored in long exposure work as the amount of time the camera is vibrating due to the mirror slamming open is miniscule compared to the overall exposure time.

Mirror lock(SCT) – Some SCT type telescopes have the ability to lock the mirror once the image is in focus to prevent the mirror from "flopping" or moving as the orientation of the telescope changes.

Monochrome – Technically means one color, meaning either black or white. "Monochrome" cameras are actually grayscale in that they produce black, white and many different shades of gray.

Mosaic - The act of shooting multiple images in a grid pattern and stitching them together to allow you to shoot a larger field of view than you could normally.

Mount - The mount is the geared (and sometimes motorized) device that is typically attached to the top of a tripod and then has the telescope attached to it. It is the mount that allows you to point the telescope at different objects without moving the tripod, and (when motorized) tracks objects across the sky.

Narrowband - Using special filters you can capture the emissions from certain gasses such as hydrogen alpha, sulpher and oxygen. These can be used much like LRGB imaging to create faux color images of high resolution. This method can also overcome all but the worst light pollution situations and can even allow you to shoot on nights with a full moon to some degree.

Near Earth Object (NEO) – An object such as a comet or asteroid which will pass in close proximity to earth.

Newtonian - A type of reflector telescope that has two mirrors in a hollow tube. The front of the telescope is open to the elements and the back is sealed. The eyepiece is near the front of the scope. These are usually not suitable for astrophotography unless they are designed as an "astrograph" as they will not bring a camera to focus without modifications or the use of a Barlow.

North celestial pole (NCP) – The point in space very close to Polaris where a line drawn from the exact southern to northern poles would extend into space with the earth revolving around that line.

Nyquist theory - States that when converting frequencies, the sampling rate should be 2x the highest frequency to get an accurate conversion and preserve all the data.

Objective lens – Also called the primary objective, the large front lens of a refractor telescope.

Off axis guider (OAG) – A method of mounting a guide camera so that it shares most of the same optical path as the imager, picking off a small amount of light usually from a mirror mounted in the light path.

One shot color (OSC) - Any camera that creates a color image from a single exposure.

Opposition – Opposition is when a planet is closest to the earth and is directly on the other side of earth from the sun.

Optical train - Anything that is directly in the path of light from the stars to your eye or camera sensor is considered "in the optical train". Could be called the optical path as well.

Optical tube assembly (OTA) – Also referred to as the OTA, this is the main tube of the telescope not including any mount, pedestal, pier or tripod.

Parfocal – Applies to both eyepieces and filters and means that if you exchange one filter (or eyepiece) for another, you will remain in nearly perfect focus. Not all filter sets or eyepiece sets are parfocal.

Periodic error (PE) - Errors in the manufacturing process of the gears and drive assembly in an EQ telescope mount results in repeating errors in the tracking of the mount. These can be removed with software that contains PEC code.

PEC - Periodic Error Correction. Software that corrects for periodic error.

Photometry – The measurement of apparent magnitude of objects such as comets, asteroids and stars.

Photon - For the purposes of discussion in this book, a photon is a single particle of light.

Photosite - The technical name for the tiny part of the sensor in a digital camera sensor that when exposed to light records a signal. Typically called a pixel.

Piggyback - Mounting a camera with a lens on a telescope in such a way as it is not shooting through the telescope but is instead just using it as a tracking mount.

Pixel - A single dot in an image.

Pixel size - The physical size of a photosite on the sensor of a camera, measured in microns.

Plate solve – Refers to Plate Solution, or finding the absolute position and motion of an object. Some applications such as TheSkyX Professional offer a plate solve feature where it can look at your image and tell you exactly what is in the frame.

Point light source - Stars are considered point light sources because regardless of their magnification they are so far away they will always appear as a single point of light.

Polar alignment – Aligning the "polar axis" of an equatorial mount to either the northern or southern celestial pole so that the mount can track celestial objects precisely.

Polar scope - A small telescope usually built into the mount which allows for precise pointing of the mount's right ascension axis to the north or south celestial pole.

Prime focus - Attaching a camera without a lens in such a way that the image from the telescope is directly projected onto the sensor of the camera.

Quantum efficiency (QE) - A measurement of the percentage of photons which hit a photosite versus how many are detected.

Rack and pinion focuser – A less expensive and typically less accurate style of focuser.

RAW - A RAW file is a file that contains the relatively unaltered, unmodified data directly from the camera's sensor.

Rayleigh scattering – The scattering of different wavelengths of light by the molecules in the atmosphere. This scattering is the reason the sky appears blue.

Resolving power – 4.56/(inches of aperture of the telescope)=resolving power of the telescope in arc-sec. Note that this does not take into consideration obstructions such as secondary mirrors.

Reticle - Crosshairs or other markings that allow you to precisely center a target in your field of view. Sometimes included inside eyepieces and finder scopes.

Red dot finder - A type of finder that uses an illuminated red dot as a reticle.

Refractor - A type of telescope that has an objective lens on the front end and an eyepiece or camera at the other. Light passes straight through without being reflected unless a diagonal is used.

RGB - Red, Green, Blue. One shot color cameras shoot everything as a combination of these three primary colors. When shooting monochrome images and wanting to end up with a color image, you shoot at least one frame with a red filter, one with a green, and one with a blue and then combine them to create a full color image.

Right ascension (RA) – Celestial coordinate measured from west to east in hours, minutes and seconds. As the earth turns each hour, 15 degrees of arc pass.

Saturation – The point at which you cannot record any more data. This may refer to the full well capacity of a CCD camera or the maximum value a pixel can store.

Schmidt Cassegrain Telescope (SCT) - a type of reflector that has a sealed front, two mirrors and has its eyepiece in the rear of the scope.

Seeing - A measurement of the conditions of the atmosphere as it relates to being able to view or image an astronomical object. An easy method to determine the seeing conditions is to look for stars twinkling; the more they twinkle, the worse the seeing.

Sidereal rate – 23 hours, 56 minutes and 4 seconds is one sidereal day which is why the stars are never at the exact same place at the exact same time every night and seem to "advance" across the night sky every night all year long. This is the rate at which your telescope must track to remain aligned with your target.

Signal to noise ratio (SNR) - The ratio of signal (what you are trying to capture in the image) to noise (electrical signals inherent to the camera generating the image). The higher the SNR, the easier it is to stretch an image and bring out the detail of your target.

Slew – The process of your telescope moving to and from targets.

South celestial pole (SCP) – The point in space very close to Sigma Octantis where a line drawn from the exact northern to southern poles would extend into space with the earth revolving around that line.

Spider vanes - Small strips of metal or plastic in the front of a Newtonian telescope which supports the secondary mirror in the optical path.

Stacking - Taking several images and combining them in such a way as to increase the signal that you want to keep while reducing the noise levels that you do not.

Strehl ratio – Gives a ratio as compared to a theoretically perfect optical system. For example, a Strehl ratio of .90 is 90% as good as a theoretically perfect optical system.

Stretching - Taking an image and manipulating the data so that details that were too dark to see are now light enough to be visible through compression of the grayscale or color scale.

T-Ring – An adapter that mates with a removable lens camera on one side and has threads on the other side to attach to the telescope or other device.

Thermo Electric Cooler (TEC) – Electric cooling device used with some CCD and DSLR cameras.

TIF - A file type (like .GIF and .JPG) to store image files. TIFs are excellent because they are lossless formats. They are however far larger than JPG or GIFs.

Tracking - The ability to follow an object as it appears to travel across the sky.

TSX - Abbreviation for TheSkyX, a planetarium, telescope control and planning application for amateur and professional use from Software Bisque Inc.

United States Naval Observatory (USNO) – The standard for timekeeping in the United States.

Vignetting - The effect of the edges of an image being darker than the center due to obstructions or optical imperfections.

Well depth - A measurement of the total amount of light a photosite can store before saturation occurs.

White point - A part of an image that represents pure white.

Zenith – The point directly overhead.

9.3: Other books by the author:

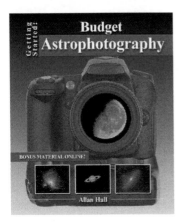

Want to take a few snapshots of the beautiful objects you are viewing without spending a small fortune? Already have a camera but you can't seem to get a good image and want to know why?

This book will answer those and many other questions while giving you a quick and reasonably easy introduction to budget astrophotography. In addition, save more money by seeing how to make a lot of items you may find useful.

http://www.allans-stuff.com/bap/

If you decide that you want more than quick snapshots, you want big beautiful prints to hang on your wall, this is the book for you.

From required and optional equipment, through the capture process and into the software processing needed to create outstanding images, this book will walk you through it all.

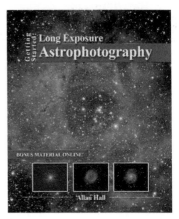

http://www.allans-stuff.com/leap/

You decide that you want to take images of celestial targets, but need a little help with the targets? This book discusses all 110 Messier targets and includes descriptions, realistic images of each target, star charts and shoot notes to help you image all 110 of the objects yourself.

http://www.allans-stuff.com/mar/

If you have ever wanted to view the wondrous objects of our solar system and beyond, here is the how-to manual to get you well on your way. From purchasing your first telescope, through setting it up and finding objects, to viewing your first galaxy, this book contains everything you need. Learn how to read star maps and navigate

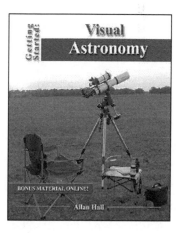

the celestial sphere and much more with plenty of pictures, diagrams and charts to make it easy. Written specifically for the novice and assuming the reader has no knowledge of astronomy makes sure that all topics are explained thoroughly from the ground up. Use this book to embark on a fantastic new hobby and learn about the universe at the same time!

http://www.allans-stuff.com/va/

Many midrange and high end telescopes come on equatorial mounts. These mounts are fantastic for tracking celestial objects. Someone who wanted to take pictures of objects in the night sky might even say they are required for all but the most basic astrophotography. The problem is that they can also be unintuitive and require some knowledge to use.

If you have ever struggled to figure out how to use an equatorial telescope mount, this is the book you always wished you had.

http://www.allans-stuff.com/eq/

So you've decided to write a book and get into non-fiction publishing. Now you find yourself faced with the seemingly infinitely harder second step – actually bringing the idea to market. In today's brave new world of self-publishing and open creative markets, it is both an inviting and potentially intimidating arena for authors hoping to turn their non-fiction books into a meaningful source of income. This is a daunting task because it

involves a blend of several disciplines that aren't necessarily part of an author's quiver of arrows. Most crucial among these are marketing and digital publishing, each of which requires fluency in fields that authors may or may not have experience in.

http://www.allans-stuff.com/ck/

WordPress is the perfect tool to help you build the website you've always wanted. But the 'help' aspect which is built into it isn't always the right thing for someone who just getting started.

What you need, and what this book will provide, is a book that shows you how to get off the ground and then build on that knowledge to give you a secure and usable website.

http://www.allans-stuff.com/wp/

Information Technology is an area which is constantly on the move, sometimes at a speed which is dizzying and difficult to keep pace with. In particular **data recovery** can be one of the more complex problems you might encounter. The sheer amount of information is often overwhelming and confusing.

Data Recovery for Normal People is a new book which aims to make this process a lot simpler. Designed for both beginners who have little knowledge of technical issues and for those who may own their own computing business and want to learn more.

http://www.allans-stuff.com/dr/

9.4: NOTES:

NOTES:

NOTES:

Made in the USA
San Bernardino, CA
29 June 2017